THE DOMESTIC CAT

GORDON STABLES

ZINC READ

Published by
This edition copyright © 2024 by Zinc Read
Design and setting By Zinc Read
Email- zincread@gmail.com
ISBN-978-93-6277-712-6

CONTENTS

CHAPTER ONE.

Classification: Its Basis.

In the feline world you find no such diversity, of form, shape, disposition, coat, size, etc, as you do in the canine. Dogs differ from each other in both the size and conformation of the skeleton, and in many other important points, almost as much as if they belonged to entirely different species. Mark, for instance, how unlike the bulldog is to the greyhound, or the Scotch toy-terrier to the English mastiff; yet, from the toy-terrier upwards to the giant Saint Bernard, they are all *dogs*, every one of them. So is the jackal, so is the fox and the wolf. The domesticated dog himself, indeed, is the best judge as to whether any given animal belongs to his own species or not. I have taken dogs to different zoological gardens, and have always found that they were ready enough to hob-nob with either jackal or fox, if the latter were only decently civil; but they will turn away with indifference, or even abhorrence, from a wild goat or sloth. But among the various breeds of cats there exists no such characteristic differences, so that in proposing a classification one almost hesitates to use the word "breed" at all, and feels inclined to search about for another and better term. If I were not under a vow not to let my imagination run riot in these papers, but to glide gently over the surface of things, rather than be erudite, philosophical, theoretical, or speculative, I should feel sorely tempted to pause here for a moment, and ask myself the question--Why are there so many distinct breeds of the domesticated dog, and, properly speaking, only one of the more humble cat? Did the former all spring from the same original stock, or are certain breeds, such as the staghound, etc, more directly descended from the wolf, the collie, Pomeranian, etc, from the fox after his kind, and other breeds from animals now entirely extinct in the wild state? And once upon a time, as the fairy books say, did flocks of wolves, foxes, wild mastiffs, and all dogs run at large in these islands, clubbing together in warlike and predatory bands, each after his kind, much in the same way that the Scottish Highlanders used to do two or three hundred years ago? Animals of the dog kind are a step or two more advanced in civilisation, if I may be allowed to use the term, than cats; and hence, as intelligence can appreciate intelligence, and always seeks to rise to a higher level, more breeds, or a larger number of species, of the former than of the latter have forsaken their wild or natural condition to attach themselves to man. May not the time come, in the distant future, when a larger variety of feline animals shall become fashionable--when domesticated tigers, tame lions, or pet ocelots shall be the rage? If so, that will indeed be the millennium for cats. Just fancy how becoming it would be to meet the lovely and accomplished Miss De Dear out walking, and leading a beautiful leopard by a slight silver chain, or Lady Bluesock in her phaeton, with a tame ocelot beside John on the dickey! A lady beside a lion on the lawn would, I think, make a prettier picture than one by the side of a peacock, and a tame Bengalese tiger would be a pet worthy to crouch at the foot of a throne. To be sure, little bits of mistakes would occur at times; instead of the pussy of the period bolting away with the canary, nothing less would satisfy the pet than a nice fat baby, and then those extraordinary people the cat--exterminators would be louder in their denunciations than ever.

If we dissect the cat, we will find that the skeleton of one breed of pussy would pretty nearly pass for that of another; we find the same shape and almost the same size of bones, the same arrangement of teeth as regards their levelness, the same number of teeth, and the same formation of jawbone. Clothe that skeleton with muscle, and still you can hardly tell the breed of the cat, for scarcely will you be able to find a muscle in the one breed that has not its fellow in all, a little difference perhaps in the size and development of one or two, but even this more the result of accident and use than a distinction real and natural.

I feel as I write that I am sailing as close to a wind as possible; I am luffing all my ship will steer; were I to keep her away a single point, I should drift down into the pleasant gulf-stream of comparative anatomy, and thence away and away to the broad enchanted ocean of speculative theory. And I confess, too, I wouldn't mind a cruise or two in those latitudes, did space and time admit of it.

Now, I do not mean to say that there is really no difference in shape and form between the different breeds of the domestic cat, but rather that this difference is so minute, compared to that which exists between dogs, that the term "breeds" seems almost a misnomer as applied to cats. It is only when you see pussy arrayed in all the wealth and beauty of her lovely fur, that you can see any real distinction between her and another.

In regard to the origin of the domestic cat, naturalists have squabbled and fought for centuries, and the best thing possible, I think, is for every man steadfastly to retain his own opinion, then everybody is sure to be right. For myself, I really cannot see that it would either assist us in breeding better cats, or render us a bit more humane in our treatment of the pretty animal, to be assured that she was first imported into this country from Egypt or Persia in the year one thousand and ever so much before Christ, or that the father of all the cats was a Scottish wild cat, captured and tamed by some old Highland witch-wife a thousand years before the birth of Noah's grandfather. What matters it to us whether the pussy that purrs on our footstool is a polecat bred bigger, or a Polar bear bred less? There she is,--

The rank is but the guinea stamp,
And a cat's a cat for a' that.

But, and if, you are fond of pedigree, why then surely it ought to satisfy you to know that, ages before your ancestors or mine could distinguish between a B and a bull, pussy was the pet of Persian princes, the idol of many a harem, and the playmate of many a juvenile Pharaoh. What classification, then, are we to make of cats? We search around us in vain for something to guide us; then, fairly on our beam-ends, are fain to clutch at the only solution to the question, and fall back upon coat and colour, with some few distinctive points of difference in the size and shape of the skull and body. Colour or markings, then, and quality of coat, are the guiding distinctions between one breed of cat and another; and to these we add, as auxiliaries, size and shape.

Colour.--Whether we understand it or not, there, undoubtedly, is nothing in this world left to chance alone, and nothing, I sincerely believe, is done by Nature without a purpose. The same merciful Providence that clothes the lambs with wool, the reason for which we can understand, paints the rose's petal, the pigeon's breast, or even the robin's egg, for reasons which to us are inscrutable, or only to be vaguely guessed at. We can tell the "why" and the "wherefore" of the rainbow's evanescent hues, but who shall investigate the laws that determine the fixed colours of the animal and vegetable creation? Who shall tell us why the grass is green, the rose is red, that bullfinch on the pear-tree so glorious in his gaudiness, and that sparrow so humble in his coat of brown?

If we ask the Christian philosopher, he will tell us that the colours in animated nature are traced by the finger of God, who always paints the coat or skin of an animal with that tint or hue, which shall tend most to the propagation and preservation of its species. That He clothes the hare and rabbit in a suit of humble brown, that they may be less easily seen by the eye of the sportsman, or their natural enemies, the polecat, weasel, white owl, or golden-headed eagle. That birds--who flit about all summer in coats so gay and jackets so gaudy, that even a hawk may mistake them for bouquets of flowers, and think them not worth eating--as soon as the breeding season is over, and the leaves and flowers fade and fall, are presented by nature with warmer but more homely suits of apparel, more akin in colour to the leafless hedgerows, or the brown of the rustling beech leaves, among which they seek shelter from the wintry blast. If you go farther you may fare worse. No one in the world can be a greater admirer than I of the genius of Tyndall, Darwin, or Huxley, but I must confess they get a little, just a *leetle*, "mixed" at times; and I doubt if Darwin himself, or any other sublunarian whatever, understands his (Darwin's) theory of colour. He says, for instance--I can't use the exact words, but can give his meaning in my own--that the wild rabbit or the hare was not painted by the finger of nature the colour we find them with any pre-defined idea of protecting the animal against its enemies; but that in the struggle for life that has been going on for aeons, considering the conditions of its surroundings, it was only the grey rabbit that had the power of continuing in existence, escaping its enemies by aid of its dusky coat. Darwin thinks, indeed, that religionists put the cart before the horse, to use a homely phrase. I confess that I myself prefer the good old theory of design--of a God of design, and a prescient Providence. I believe the testimony of the rocks, I believe to a great extent in evolution--it is a grand theory, and one which gives the Creator an immensity of glory--but I cannot let any one rob me of the belief that beauty and colour are not all chance.

Yonder is a hornet, just alighted at the foot of the old oak-tree where I am writing, so uncomfortably near my nose, indeed, that I can't help wishing he had kept to his nest for another month; but the same April sunshine that lured me out of doors lured the hornet, and there it stands, all a-quiver with delight, on a budding acorn, looking every moment as if it would part amidships. "Do you think, Mrs Hornet, O thou tigress of bees, if your lovely body, with its bars of gold, had been of any other colour, that, under the peculiar conditions in which your ancestors lived, you would, ages ago, have ceased to exist; that ants, or other 'crawling ferlies,' who detest the colour of turmeric, would, in spite of your ugly sting, have devoured you and yours?"

Yonder, again, is a beautiful chaffinch; he was very glad to come to my lawn-window every day, during all the weary winter, to beg a crumb of bread. He forgets that now, or thinks perhaps that I do not know him in his spring suit of clothes, and golden-braided coat and vest. But I do, and

I still believe--simple though the belief may be--that the same Being, who gave life and motion to that little beetle which is now making its way to the highest pinnacle of my note-book, as proud as a boy with a new kite, to try its wings for the first time, tipped that ungrateful finch's feathers with crimson, white, and gold, in order to make him more attractive to his little dowdy thing of a wife, who has been so busy all the morning building her nest on the silver birch, and trying to find lichens to match the colour of the tree. For Mrs Finch is a nervous, timid little body, and had no thoughts of marrying at all, and indeed would have preferred to remain single, and would have so remained, had she not been a female; but being a female, how could she resist that splendid uniform?

I go into the garden and bend me over the crocus beds--white crocuses, orange crocuses, and blue, all smiling in the sunshine of spring. Each is a little family in itself, and they would like to know each other too so very much, for they have ever so many love tales to breathe into each other's ears. But they are all fast by one end and cannot move. Whatever shall they do, and what will become of the next generation of crocuses? I can hear them whispering their tales of love to the passing wind, and so can you if you are a lover of Nature; but the wind is too busy, or too light, or too something or another, and cannot pause to listen. So the little things are all in despair, when past comes a bee. Now bees, and butterflies too, for all they have got so many eyes, are rather short-sighted, but even a bee cannot help seeing that gorgeous display of orange, white, and blue, so he pops at once into the bosom of a blue crocus, and is made as welcome as the flowers in May.

"Oh! you dear old bee," says the crocus, "you're just come in time; have something to eat first. I have a nice little store of honey for you; and then you shall bear a message to my lady-love--the pretty blue belle crocus mind, not the white. I wouldn't have a race of variegated children for the world."

"All right," says the bee, and away he flies with the message of love to the blue belle crocus, and thus the loves of the crocuses are cemented. They tell the old, old story by proxy, because they can't do it as you or I do, reader, eye to eye and lip to lip.

For colour has its uses, and nothing that exists was made in vain, although some are selfish enough to believe that all the colour and beauty they see around them, during a ramble in the country, was made but to please the eye of man.

Colour I believe to be connected in some way with the mystery of heat and life. We all know that certain colours will dispel or retain heat; black is more warm, for instance, than white. There may be, then, a *scale* of colours as it were, each colour differing in the amount of heat-retaining power; and, it may be that, having reference to this scale, the colours on an animal's coat, are apportioned to it in the way which shall best conduce to its health, comfort, and happiness.

The colour of any animal is an important consideration in determining its breed, and this is especially the case among cats, where indeed it forms the basis of our classification. Colour is often the key to the character of the cat--to its temper, whether savage or good-natured; to its qualities as a good hunter or the reverse; and to its power of endurance, its eyesight, and its hearing.

Size.--Cats of different breeds--I use the word for want of a better--are generally of different sizes, and the skeleton is, as a rule, larger in some breeds than in others. The male ought to be larger than the female.

Form.--The difference in form is principally observable in the shape and rotundity of skull, the length and shape of the nasal bones and jaw, and the length of the tail and its form at the point. The ears also vary a good deal in length in the different breeds, and also in breadth, and in "sit" or position.

Pelage, or Coat.--The coat is of two different kinds, the long and the short. In the former, the longer and softer and silkier the better, and in the latter the length of the hairs, their closeness and glossiness, are to be taken into consideration. You can generally tell by one glance at the animal's coat how she is fed, how she is treated and housed, and the condition of her health.

Having got so far, we will next bring pussy herself on the stage, and see how far these remarks apply to her, according to her breed and species.

* * * * *

CHAPTER TWO.

Breeds and Classes.

In future chapters I will give the habits and characteristics of the domestic cat in general, with some specialities of a few of the different kinds in particular. The "tricks and manners" of one cat, however, will be found to correspond pretty closely with those of any other.

But before going farther on with this chapter, I wish to make a plea in pussy's favour. I myself have studied cat life, off and on, for twenty years, so I suppose it will be admitted I am no mean authority on the subject. During that time I have come to certain conclusions, which in some cases run contrary to the opinions generally conceived of those animals--contrary, at any rate, to the belief current some years ago, before pussy was thought worthy to hold a show of her own. Towards this ocean of contrary opinions I have been wafted, not by the wind of my own sails alone, but aided and supported by many hundreds of anecdotes of domestic pussy's daily life, habits, likes and dislikes. These anecdotes have been supplied to me from trustworthy people, in every position of life--from the poverty-stricken old maid with her one feline favourite; from the honest working-man with his fireside pet and children's playmate; from farmers, solicitors, doctors, and parsons; from baronets' ladies; and, in more than one instance, from the daughters of peers of the realm, allied to royalty itself. These anecdotes have, in almost every case, been substantially authenticated, and *always* discarded wherever, in any case, they were open to doubt.

From these anecdotes and essays, and from my own experience as well, I have arrived at the following conclusions--and be it remembered I speak of cats that are properly fed and housed, and have been taught habits of cleanliness when kittens:--

1. That cats are extremely sagacious.

2 That cats are cleanly and regular in their habits.

3. That cats are fond of children.

4. That cats are excellent mothers, and will nurse the young of any small animal on the loss of their own.

5. That cats are fond of roaming abroad.

6. That cats are brave to a fault.

7. That cats are fond of other animals as playmates.

8. That cats are easily taught tricks.

9. That cats are excellent hunters.
10. That cats are good fishers, and can swim on occasion.
11. That cats are very tenacious of life.
12. That cats are fond of home.
13. That cats are *fonder far of master or mistress.*
14. That cats are *not*, as a rule, *thieves, but the reverse.*
15. That long-headed, sharp-nosed cats are the best mousers.

These are not texts, but deductions.

All that is known for certain of the origin of the domestic cat may be expressed in three letters, *n i l*--nil. And, after all, I cannot see that it matters very much, for if the theory of Darwin be correct, that everything living sprang originally from the primordial cell, then cats or dogs, or human beings, we all had the same origin. But, again, according to Darwin, the cat is an older animal than man in the world's history; and if this be so, how silly of us to bother our heads in trying to find out who first domesticated the cat, when in all probability *it was the cat who first domesticated man*. But, avaunt! all learned discourse on the subject; perish all discursive lore. I have studied the matter over and over again, and read about it in languages dead and living, till my head ached, and my heart was sick; and still, for the life of me, I cannot make out that there are any more than two distinct *species* of domestic cats in existence. There are, first, the European or Western cat, a short-haired animal; and secondly, the Asiatic or Eastern cat--called also Persian or Angora, according to the difference in the texture of the coat, it being exceedingly fine, soft, and satiny in the Angora, and not so much so in the Persian--a long-haired cat. All the others, such as Assyrian, Abyssinian, the Maltese, Russian, Chinese, Italian, French, Turkish, etc, are either inter-breeds between the two, or lineal descendants of the one or the other, altered and modified by climate and mode of life.

Taking everything into consideration, I am inclined to favour the belief held by some, that our own fireside cat was first domesticated from our mountain wild cat. I mentioned, this to a naturalist of some repute, with whom I was dining only a few days ago.

"*What*?" he roared, trying to get across the table, in order to jump down my throat. "*You* ought to know, sir, that all animals increase, instead of degenerating in size, by being transplanted to domestic life."

I didn't contradict the man in his own house; but indeed, reader, the rule, if rule it be, admits of numerous exceptions. It holds good among horses, and I suppose cattle of all kinds; it even holds good if we go down the scale of organic life, and apply it to fruit and flowers; but how about the wilder animals, and our forest trees? Take the latter first--will the acorns of a garden-grown oak-tree, or the cone of a transplanted Scotch pine, produce such noble specimens as those that toss their giant arms in the forest or on mountain-side? Or will a menagerie-bred lion, or tiger--feed them ever so well--ever reach the noble proportions of those animals who in freedom tread the African desert, or roam uncaged and untrammelled through the jungles of Eastern India? What prison-born elephant ever reached in height to the shoulders even, of the gigantic bulls that my poor friend, Gordon Cumming, used to slay? Do eagles, owls, the wilder hawks, alligators, or anacondas do anything else but degenerate in captivity? But even admitting, hypothetically, that the rule would hold good as regards cats, there isn't such a very great difference in the size of the tame and wild cats after all. I do not think that all the wild cats ever I saw in Scotland or elsewhere, would average over ten to twelve pounds; and twelve pounds is no unusual weight for our domestic cheety. Another thing that has often struck me is this: the farther north you go in Scotland, and the nearer to the abode of the wild cat, the greater is the resemblance in head and tail, and often in colour, of the tame cat to the wild. And, mark you, the domestic is often known to inter-breed with the wild cat, and the offspring can be tamed and reared. This is considered nothing unusual in the Highlands.

* * * * *

CHAPTER THREE.

Breeds and Classes. The Tortoiseshell.

The classification I propose of the domestic cat is an exceedingly simple one, as I think all classifications ought to be; it will, I trust, however, be found quite sufficient, and a useful one. We have first, then, the two and only two distinct breeds mentioned above, viz:--One. The European Cat. Two. The Asiatic.

From these two alone, if you get them of different colours, you can very easily manufacture all the varieties and various-coloured pussies you are ever likely to meet with, either on the show-benches or in domestic life.

One. The European, short-haired, or Western Cats.

These I divide into five primary classes, namely--1, *Tortoiseshell*; 2, *Black*; 3, *White*; 4, *Blue* or *Slate-colour*; and 5, the *Tabbies*.

The *Tortoiseshell* I subdivide into secondary classes: 1, the pure Tortoiseshell; and 2, the Tortoiseshell-and-white.

The *Black* is subdivided likewise into two: 1, pure Black; and 2, Black-and-white.

The *White* has no subdivision, but is bred in with any or all the other classes.

The *Blue* or *Slate-coloured Cat*. These are subdivided into two: 1, the pure Blue; and 2, the Blue-and-white.

Tabbies are easily subdivided into four classes, viz:--1, the Red Tabby; 2, the Brown Tabby; 3, the Blue or Silver Tabby; and 4, the Spotted Tabby.

There are other odd cats, such as the Manx or tailless cat, the hybrid, the six-clawed cat, and some curiously-coloured animals, which I shall mention in another place, for these have no right to have classes of their own, any more than black-and-tan Newfoundlands, or kittens with eight legs.

I shall take these in their order of rotation.

1. *The Tortoiseshell Cat*.--This might also be called the black-and-tan cat. If you want to get a good idea of the colour this cat is, or ought to be,

take a keek through a lady's tortoiseshell back-hair comb. That is about it; but you never see such perfection in pussy's coat.

For many a long year it was almost universally believed that there never was any such thing as a tortoiseshell male or Tom cat, or ever could be; and many an anxious search has many an old maid had over her newly-born litter of kits, to see if she would be fortunate enough to find the much-to-be-desired anomaly. For, bear in mind, a belief used to be pretty current that 300 pounds--or was it 500 pounds?--would be paid over some counter, by some fool or fools unknown, to anyone who should be able to put the possibility of the existence of a tortoiseshell Tom beyond dispute--by producing one. I saw an advertisement the other day in *The Live Stock Journal*, offering for sale a tortoiseshell Tom, at the low price of 100 pounds! I hope, if only for poor Tom's sake, that somebody with more money than brains bought it--for the cat anyone paid 100 pounds for would, I should think, be certain of good milk and generous treatment.

I knew a poor old woman in Skye, and this old woman's pussy was as pussies love to be. And lo! one night the old woman, in the silence of night, dreamed a dream. She thought that the cat came to her bedside, and said to her, "Arise, mistress, come and see." That she followed pussy at once. That pussy led her to the barn. That there she found, cuddled together in a heap upon an old sack, no less than five tortoiseshell Toms. She dreamt besides that she sold the lot for 1,000 pounds each, and bought a carriage and four, right off the reel, and set up for a lady of fashion on the spot. Anxiously did the old woman watch for her cattie's accouchement, but much to her disappointment they all had white about them. Next time that pussy was in the same way, her mistress had an old tortoiseshell comb nailed up above its bed. Even this didn't do, so--for by this time the ancient dame had tortoiseshell Tom on the brain--she set out for Portree, a distance of fully sixteen miles, where she managed to procure a live tortoise as a playmate for her pet. Pussy never took much to the tortoise; all she did was to sit and watch it, and whenever it protruded its scaly head, the cat smacked it in again. This might have been the reason why her kittens had all white about them the third time. The old woman didn't despair, however; she took to praying, and prayed in English, and prayed in Gaelic, and she told me seriously that she never doubted but that her prayers would one day be answered--*if,* she added, *it was for her good*. I didn't doubt it either, but Tom never came ashore as long as I was in the island, neither was the old creature's snuff-box ever empty.

I have but little fancy for this breed myself. They are usually sour-tempered, unfriendly little things to all save those who own and love them. They are, moreover, not very prepossessing. I speak of the cat as *I* have found it, and I doubt not there are many exceptions.

Merits.--They are excellent and patient mousers, and the *best* of hunters. They are likewise good mothers. They are as game as a bull and terrier--in fact they seem to fear nothing on four legs; and when they do take off the gloves to fight, I pity the animal they tackle, for what the tortoiseshell lacks in weight, she makes up for amply in courage. They are very wise and sagacious, and faithful to the death to those who own them.

Points.--1. *Size*: You don't look for a very large cat of the pure tortoiseshell breed, nor a very pretty one. The larger the better to a certain extent. I have known a small-sized tortoiseshell cat follow the rats even into their own burrows, again and again, until she had exterminated them. 2. *Head*: The head is small and rather bullety, the ears moderately large and nicely cocked, and the eyes small, and the darker the better. 3. *Colour and markings*: The colour is as near tortoiseshell as possible, and the markings must not only be deep and pretty, but very distinct in the centre, although blending insensibly where they meet, and artistically arranged. You mustn't expect to find the colour or markings very nicely arranged on the male tortoiseshell. No white is allowed on this breed of cat. Tortoiseshell Tom *is* tortoiseshell Tom, and prefers to be judged alone and on his own merits; for, as a rule, his right there is none to dispute. 4. *Pelage* or *Coat*: Hair moderately short, but *very* fine, glossy, and silken. N.B.--Knock off from five to eight points for *cinder-holes*. I now give the points in a tabular form, with their full value. Not, remember, that as a rule I go in for judging by points; still, a table of this sort has its value, as one can see just at a glance what is looked for in each breed, and what isn't:--

---

Points of the Tortoiseshell Cat.

1. Size, 5.
2. Head, 10.
3. Colour and markings, 25.
4. Pelage, 10.

Total, 50.

The next pussy which demands a few passing remarks is *The Tortoiseshell-and-white*. This is often a very beautiful cat, more especially when young, as, when old, they sometimes degenerate into very lazy habits, especially if they have a large amount of white about them. They are pretty, and they seem to know it, taking great delight in keeping the white portions of their fur as pure as snow. I knew a cat something of this breed, who was nearly all white, excepting a beautiful tortoiseshell patch on the upper part of one thigh. She was unexceptionably cleanly, and the frantic efforts she used to make to wash off that spot of black-and-amber were ridiculous to behold. She would sit for hours admiring herself in the glass, and occasionally dipping her paw in her saucer of milk, until she spied that unhappy spot; to that she would at once devote a good half-hour, but finding no appreciable difference in it, she would start away in high dudgeon, swishing her tail about, like a lion in love. That spot was the only barrier to pussy's bliss. *Moral*: There's no such thing as perfect happiness here below--even to a cat.

* * * * *

CHAPTER FOUR.

The Black Cat.

Next on the list of classification comes the *Black Cat*, subdivided into--1, the Pure Black; and 2, the Black-and-white.

1. *The Pure Black*.--This is one of my pet breeds. The pure black cat is such a noble, gentlemanly fellow, and if well-bred and trained--and he is capable of a very large amount of training--he is one of the best and most useful cats you can have in the house. There is no namby-pambiness about black Tom, and no squeamishness either. You can take him or tire of him, just as you please; it is all one to Tom. There is a certain independence about his every movement, and an assumption of dignity, as he saunters about the house, gazes at the fire of a winter's evening, or rolls himself in the sunniest spot of the garden in summer, that are both amusing and delightful. Black Tom will give you a paw, but you may take it or leave it, just as suits you; and if you annoy him too much, he will very quickly cast his gloves and make you laugh with the wrong side of your mouth, as the saying is. And it is quite astonishing, too, what a beautiful deep and cleanly-cut wound--I speak feelingly, as a surgeon--Tom can make on the fleshy portion of your hand, or down the side of your nose. For black Tom, and all the race of black cats, seem to have made up their minds ages ago not to stand any nonsense from man or beast.

But you mustn't run away with the idea that black Tom is a pugnacious animal, or fond of fighting for fighting's sake. No, Tom is never aggressive; he stands a good deal before he is thoroughly roused, and, to tell the truth, I have more than once seen a tortoiseshell thrash a black cat double its size. But if there is a lady cat in the play, the affections of a queen to be gained, or if black Tom has made up his mind to carry war into the heart of a rival's camp, doesn't he go at it with a will! If the other cat will not surrender, ten to one all you'll find of that cat in the morning will be the front teeth, the wind of the battle having blown all the fluff away, while, if you cast your eyes upwards, you will see black Tom on the top of the wall making love to his Dinah, and looking as if butter wouldn't melt in his mouth.

Black Tom is generally most exemplary in the matter of cleanliness, personal or otherwise--there you have him again. And he is as proud as Lucifer--for he is quite well aware that he *is* good-looking. If he were a man, he is just the sort of fellow who would wear a well-fitting coat, spotless linen, and well-fitting boots and gloves, and part his hair in the centre without appearing a cad. You will seldom see cinder-holes in black Tom; if you do, you may lay your honour on it, that the animal is either aged and infirm, or suffering from some internal disorder.

The black cat might be called the Newfoundland of the feline race, not only in colour, but in nearly all his ways. He is not the pussy, however, I like to see made a pet of by children, for two reasons--first, he is too fine an animal to be crumpled and spoiled; and, secondly, because, like a good many Newfoundlands, he is liable at times to be just a little uncertain in temper.

Although he cannot save life, like his prototype, still black Tom makes the best of black guards, and will protect his master or mistress, or their property. One or two that I happen to think of now, keep a watch on their master's wares just as a dog would. One belonging to Mr Taylor, of Cumministon, "clooked" a little boy in the very act of stealing a piece of butter, and held him, growling fiercely the while, until his master came. The same cat would keep the packet of groceries ordered by a customer, until the money was paid, and he was told it was all right. The cunning and wiliness of the black cat is sometimes highly amusing. I have known a cat of this breed feign death to escape a thrashing; that is, when being thrashed, he pretended that one of the blows had suddenly killed him, and would lie to all appearance stark and stiff on the floor for several minutes; but if you watched him narrowly you would presently see just a line of his cute brown eye, and as soon as the coast was clear, Tom would come to life again, and be off like a shot.

Black cats are sometimes thieves. I know the reader would put it in more forcible language, but don't you expect for a single moment that I will say more against my pets than the exigencies of truth compel me to, so there! I say they are at times just a *leetle* addicted to appropriating what they have but small legal right to. But there is this to be said in their favour--when they are thieves *they are swells at it*. I have a black cat in my eye at this very moment, and if, my *dear* lady, you are at all fond of that sort of thing, it would, simply do your heart good to watch that pussy stalking steak. He is such an honest-looking cat, you see, and from the easy way he sits in the doorway opposite the butcher's, with his half-shut eyes and his dreamy air, you would feel convinced that the house was his home, that all the adjoining property belonged to him, and he had a vote in Parliament and a seat on the municipal bench. But bide a wee till Blocks turns round to serve a customer, when pop! fuss!! honest Tom is round the corner with a pound of beef in his mouth, before you could say "Muslin!" Oh! it's charming, I assure you, but rather rough on Blocks.

I must confess, too, that, at times, there is about a black Tom cat a look which you can only designate as Satanical--Mephistophelean, then, if you object to the other word--and I have no doubt it is this look of devil-beauty in Tom which has often led him to be suspected of being either an imp of darkness or possessed of one. A witch, you know, is generally supposed to have as a companion a familiar spirit in the shape of a black cat. Superstitions connected with the black cat are still common in some parts of the country and among sailors. We had a black Tom in the *Penguin* which led us many a pretty dance. He was treated as a fiend, poor fellow, and behaved as such; and the captain was as much afraid of him as anybody else, and never failed to let go the life-buoy and lower a boat when Tom missed his footing and fell overboard, which the cat had a happy knack of doing periodically. Tom was missed, though, one morning, and seen again no more. He had doubtless fallen into the sea in the darkness of the middle watch.

This cat had a strange method of fishing, which is worthy of notice. You are, I suppose, aware that flying fish are caught by exposing a light on deck, which they always vault towards. Black Tom's eyes had the same effect. He would sit on the bulwarks and glare into the sea till a fish flew towards or over him, then he nabbed it nimbly. Just before we came to the Cape, for the last time in that commission, I heard two blue-jackets conversing about this black Tom.

"Look, see!" one was saying, "I think he were a devil, nothin' more and nothin' less; and I'll bet you five bob he were a devil."

"Done," said the other sailor.

Three days after, both men were "planked" for coming off drunk. They had been on shore drinking their bet beforehand. Simple souls, they both came to me after punishment, to get my decision as to who should pay. Their doctor, they thought, knew everything. But very sadly were they put out, when I told them the bet could never be satisfactorily decided *in this world*.

"Ah! doctor," said one, waggishly, "it's a jolly good thing we drank the bet beforehand."

Black Tom's queen is usually a very lively lady, and up to any amount of fun and mischief.

Merits.--For house-hunting they are the *best* cats you can have. They are very beautiful and graceful; and, indeed, a well-bred, well-trained black Tom is a veritable prince of the feline race. The finest cat of this sort

I ever saw was at Glasgow Show, "Le Diable" to name. He *was* a beauty. What attitudes he did! What grace in every movement! and such a colour and coat and eye! I forget now who owned him, but I remember I gave him first prize after only one glance at the others. Black cats are not so easily seen at night, and their hearing is extremely keen; so, likewise, is their eyesight. As a rule, they kill rats and mice more for sport than anything else, and are fonder of tackling larger game. In the field, however, their colour is against them, and makes them a good mark for the keeper's gun. I prefer seeing black Tom in the parlour, or on a hosier's counter, or coiled up in a draper's window.

Points.--1. *Size*: You want them large--as large an possible, and with great grace of motion. 2. *Head*: The head is medium-sized, and not too bullety; a sharp nose, however, is an abomination in a black cat. The ears must be rather longish, and shapely, and well-feathered internally, and set *straight* on. 3. *Eyes*: A brown eye is best, next best is hazel, which in turn is better than green, but green is better than yellow. 4. *Colour*: All black; not even a toe must be white, nor *one hair of the whiskers*. 5. *Pelage*: A beautiful, soft, though not too fine, fur, and inclining rather to length than otherwise, and as sheeny as a boatman beetle.

Points of the black cat.

Size, 15.

Head, 5.

Eyes, 5.

Colour, 15.

Pelage, 10.

Total, 50.

* * * * *

CHAPTER FIVE.

The Black-and-White Cat and the Pure White.

I have been asked to give a few hints as to the best and most useful classification for show purposes, and may as well do so here. For a large show, the classes can hardly be better arranged than they are in the Crystal Palace catalogue, or that of the Edinburgh or Glasgow Shows. For smaller shows I beg to suggest the following:--

One. Long-haired cats, any colour, male or female.

Two. Short-haired black and black-and-white, and white.

Three. Short-haired tabbies, any colour.

Four. Short-haired tortoiseshell and tortoise-and-white.

Five. Anomalous, as Manx, *etc.*

The first class would include Persian, Angora, and other long-haired cats--black, white, tabby, or tortoiseshell. The third class would include all tabbies--brown, red, and grey or silver. Class Four must have tortoiseshell-and-white as well as tortoiseshell, or it will be a small class, owing to the rarity of the pure tortoiseshell. The last class will give a place to Manx, six-toed cats, wild cats, and hybrids, as well as any curious foreign pussy that may be forthcoming. At all shows you find a great many cats entered in the wrong class. I think it a pity that secretaries don't arrange these in their proper classes; it is not right to exclude merit through mistake. In judging, prizes should be withheld where there is no competition; and where there is want of merit in any one class, some of the prizes should be withdrawn and added to any class of *extra* merit. We come now to the *black-and-white cat*.

A good black-and-white cat is a very noble-looking animal. If well-trained and looked after, you can hardly have a nicer parlour pet. He is affectionate in his disposition, and cleanly and gentlemanly, so to speak, and makes himself quite an ornament to a well-furnished drawing-room. I must speak, however, of the demerits of my pets, as well as of their good qualities, and feel constrained to say that I have sometimes found black-and-white Tom a pussy who did not trouble himself too much about his duties as house-cat; he much preferred the parlour to the kitchen, a good bed to a hay-loft, and seemed to think that catching mean little mice was far below his dignity. If well treated black and white cats are apt to turn a little indolent and lazy, and if improperly fed and housed, they degenerate into the most wretched-looking specimens of felinity you ever looked upon. All the bad in their character comes out, and their good qualities are forgotten. Their coat gets dry, and tear, and are cinder-holed; and, instead of the plump, round-faced, clerical-looking cat which used to adorn your parlour window, you have a thin, emaciated, long-nosed, pigeon-loft-hunting, flower-unscraping, dirty, disreputable dunghill cat. Of course, the same may, to a certain extent, be said of most neglected cats, but the two breeds that show to the least advantage, when ill-used, are the black-and-white and the red-and-white, and more especially the former.

Merits.--I like these cats more for their appearance than anything else. When nicely marked they look reverend and respectable in the extreme. I consider them but very ordinary pussies in regard to house-hunting. A naval officer who cannot go to quarters without having his hands encased in white kids, and a black-and-white cat, carry on duty much on a par. Neither do these cats make over good children's pets, being at times a little selfish. They are beautiful creatures, nevertheless, and well worthy of a place at our parlour firesides.

Points.--1. *Size*: As big as possible, but not leggy; reasonably plump for the show-bench, but *very* graceful in all their motions; with stoutish short forelegs, and plenty of spring in the hindquarters.

2. *Head*: The best black-and-white Toms have large, well-rounded heads, with moderately long ears, and a well-pleased, self-contented expression of face. The whiskers are usually white, but black is not objectionable. The eyes are preferred green, and sparkling like emeralds of the finest water.

3. *Colour and markings*: The colour is black-and-white, with as much of the former, and as little of the latter as you can find. I like to see the nose and cheeks vandyked with white, the chin black, white fore-paws, white hind legs and belly, and a white chest. This is all that is needed for beauty's sake; but, at all events, the markings must be even.

4. *Pelage*: Fur should be longish (and I don't object to its being ticked all over the back with longer white hairs), silky, and glossy.

Points of the Black-and-White Cat.

Size, 10.

Head, 5.

Colour and markings, 25.

Pelage, 10.

Total, 50.

The next cat on the boards is the white cat.

It is very remarkable--and most students of feline nature must have had an opportunity of observing this--the great difference in the temperament, constitution, and nature of cats, which colour alone, apparently, has the power of truly indicating; and this is nowhere more easily seen than in the peculiar characteristics of the pure black pussy and the all-white one. The black cat, on the one hand, is bold, and free, and fierce; the white, far from brave, more fond of petting and society, and as gentle as a little white mouse. The black cat is full of life and daring; the white of a much quieter and more loving disposition. The black cat stands but little "cuddling;" the white would like to be always nursed. It takes but little pains to teach a black cat to be perfectly cleanly, but much more to train a pure white one. In constitution the black cat is much more hardy and lasting, the white cat being often delicate, and longing apparently for a sunnier clime. A black cat is often afflicted with *kleptomania,* while a properly-educated white puss is as honest as the day is long.

The senses of the black cat are nearly always in a state of perfection, while the white is often deaf, and at times a little blind. Again there is nothing demoniacal about a white cat, as there often is about a black one. I remember, when a little boy at the grammar school of Aberdeen, receiving a box from the country containing lots of good things, and marked, "A Present from Muffle"--Muffle was a pet tabby of mine--and, childlike, replying in verse, the last lines of the "poem" being--

"And when at last Death's withering arms
Shall throw his mantle thee around,
May angel catties carry thee
To the happy hunting-ground."

Well, a blue-eyed white pussy was my idea of an "angel cattie" then, and it is not altered still.

It will be observed, however, that the colour of the kittens of the same litter will often differ, and the question naturally comes to be asked, Do I assert that the nature and temperament of cats in the same litter will not coincide? I do so aver most unhesitatingly; and the thing is easily explained if you bear in mind that a litter of differently-coloured kittens has had but *one* mother, but *many* fathers. Although born from the same mother in one day, they stand in the relation to each other of half-brothers and half-sisters. Except when the odds in colour is very distinct, as in black, white, or red, the difference in constitution, etc, will not be so easily perceived, but it is there, nevertheless. *Colour follows the breed, and temper and quality follow colour*. This is the same all throughout nature, and is often observed, though but little studied, by dog fanciers. I have only to remind pointer and setter men, how often hardiness and good stamp cling to certain colours. That "God tempers the wind to the shorn lamb," I believe to be merely metaphor, but I am ready to go to death on it that He paints the petals of the flower and the blossoms on the fruit-tree, to the requirements of the tender seedlings. What sort of fruit would you grow in the dark, or under deeply-coloured glass shades? Lest I be found guilty of digression, I shall say no more now on this subject.

Merits of the White Cat.--A *pet*, gentle and loving above a cat of any other colour, though at times dull, and cross, and wayward; "given," as a lady said, "to moods of melancholy." Not a bad mouser either, when "i' th' vein," and a good cat for a miller to have, not being easily seen among sacks of flour.

Points.--1. *Size*: Seldom a large cat. 2. *Head*: Smallish, and as nicely rounded as possible; ears not too long, and well-feathered internally; eyes of "himmel-blue;" eyes ought to be both the same colour--if not so, deduct five points. 3. *Pelage*: Fine, soft, and glossy; but a too long coat shows a cross with Angora. 4. *Colour*: White as driven snow, if intended for a show cat; if not, a very little black wonderfully improves the constitution.

Points of the White Cat.

Size, 5.

Head and eyes, 15.

Colour, 25.

Pelage, 5.

Total, 50.

* * * * *

CHAPTER SIX.

The Blue Cat; and Tabbies--Red, Brown, Spotted, and Silver.

The Blue cat: just one word about this pretty creature before passing on to the Tabbies. Although she is called a blue cat, don't fancy for a moment that ultramarine is anywhere near her colour, or himmel-blue, or honest navy serge itself. Her colour is a sad slate-colour; I cannot get any nearer to it than that.

Apart from her somewhat sombre appearance, this cat makes a very nice pet indeed; she is exceedingly gentle and winning, and I'm sure would do anything rather than scratch a child. But the less children have to do with her the better, for all that: for this simple reason--she is a cat of delicate constitution--all that ever I knew were so, at least, and I daresay my readers can corroborate what I say.

Merits.--Their extreme gentleness is one merit, and their tractability and teachability are others. A pure blue cat is very rare, and they are greatly prized by their owners.

Points.--1. *Size*: They are rather under-sized, never being much larger than the pure tortoiseshell.

2. *Head*: The head is small and round, and the eyes are prettiest when of a beautiful orange-yellow. The nose should be tipped with black.

3. *Pelage*: Moderately long and delightfully soft and sheeny.

4. *Colour*: This is the principal point. It is, as I said, a nice cool, slate-grey, and, like the black cat, our blue pussy must be all one colour, without a hair of white anywhere. *Even her whiskers* must be of the same colour as her fur.

Points of the Blue Cat.

Size, 5.

Head, 5.

Pelage, 10.

Colour, 30.

Total, 50.

We now come to the Tabbies--the real old English cats--the playmates of our infant days and sharers of our oatmeal porridge. They are the commonest of all cats, and justly so, too, for there is hardly anything they don't know, and nothing they can't be taught, bar conic sections, perhaps, the *Pons Asinorum*, and a few trifles of that ilk. You will find a tabby cat wherever you go, and you will find her equally at home wherever she is--whether sitting on the footstool on the cosy hearthrug, singing duets with the tea-kettle; catching birds and rabbits in the woods, or mice in the barn; conducting a concert for your especial benefit on the neighbouring tiles at twelve o'clock at night; examining the flower seeds you lately sowed in the garden to see if they are budding yet; or locked, quite by accident, into the pigeon loft.

The first cat of the Tabby kind which claims our attention is the Red or Sandy Tabby.

This is a very beautiful animal, and quite worthy of a place in the best drawing-rooms in the land. Although they do not grow to the immense size of some of our brown tabbies, still they are better hunters, much fiercer, and of a hardier constitution. They much prefer out-of-door sport, and will attack and slay even the polecat and weasel; and instances have been known of their giving battle to the wild cat himself.

Merits.--They are the prettiest of pets, and the honestest of all cat kind. They are such good ratters that neither mice nor rats will frequent the house they inhabit.

Points.--1. *Size*: They ought to be as large as possible, and not clumsy; they are generally neater cats all over than the Brown Tabbies.

2. *Head*: The head should be large and broad, with rather shortish ears, well placed, and the face ought to beam with intelligence and good nature. The eyes should be deep set, and a nice yellow colour.

3. *Pelage*: The coat is generally short in nearly all the Tabbies, but ought to be sleek and glossy.

4. *Colour and markings*: The colour is a light sandy red, barred and striped with red of a darker, deeper hue. No white. The stripes or markings ought to be the same on both sides, and even the legs ought to be marked with cross bars, and one beautiful swirl, at least, across the chest. This is called the Lord Mayor's Chain, and when the cat has two, give him extra points.

Points of the Red Tabby.

Size, 10.

Head, 5.

Colour and markings, 30.

Pelage, 5.

Total, 50.

Next comes the Brown Tabby.

This is the largest of all breeds of cats, fourteen, seventeen, and even twenty pounds a common weight. They are also, when well marked and striped, exceedingly beautiful. Of all cats they are the best adapted for house-hunting, being less addicted to wandering than some breeds.

Merits.--Their hunting proclivities. Their fondness for children is sometimes quite remarkable. I have known many instances of Brown Tom Tabbies, so fierce that scarce any one dare lay a finger on them unscathed, but a little child of four years of age could do anything with them, lug them about anyhow, and even carry them head down, over its shoulder by the tail. They are, moreover, nice, loving, kind-hearted pets, and exceedingly fond of their master and mistress. They are the cats of all cats to make a family circle look cosy and complete around the fire of a winter's night.

Points.--1. *Size*: It will be observed below that I give fifteen points for size. The bigger your Brown Tom Tabby is the better he looks, *if* the one-half of it isn't fat, for if so he won't be graceful, and that is one essential point. I can find a Tabby at this moment who weighs over twenty pounds, and who will spring from the floor, without scrambling, mind you, clean on to the top of the parlour door, and that is little short of seven feet. I like to see a tabby with a graceful carriage then, and shortish in forelegs, with beautifully well-fitted and rounded limbs, and with a tiger-like walk and mien.

2. *Head*: Very large and broad and round, ears short, eyes dark, and muzzle broad, not lean, and thin and long. This latter certainly gives him more killing power, but it brings him too near the wild cat. I don't care how savagely he behaves in a cage at a show, for well I know he is quite a different animal at his own fireside, asleep on the rug in little Alice's arms, or purring in bed on old Maid Mudge's virgin bosom.

3. *Colour*: A nice dark brown or grey ground, and the workings as deeply black as possible. No white.

4. *Markings*: Like a Bengal tiger, and even prettier. The tail and legs likewise barred. The head striped perpendicularly down the brow, and the marks going swirling round the cheeks. Nose black or brown, and the eyes as dark as possible, and full of fire.

5. *Pelage*: Short and glossy.

Points of the Brown Tabby.

Size, 15.

Head, 5.

Colour, 10.

Markings, 15.

Pelage, 5.

Total, 50.

Lastly, we have the Silver Tabby and the Spotted Tabby, and in almost all points these may be judged alike.

The Silver Tabby is a sweetly pretty cat. Perhaps the prettiest of all pussies. They are a size smaller than even the best Red Tabbies, and are infinitely more graceful, and quicker in all their motions. They are proud, elegant, aristocratic cats, fond to love and quick to resent an injury.

Merits.--Their special merit is their exceeding beauty. They are somewhat rare, however. Here is a bit of advice to any one who would like to have four really pretty cats about the house, each to show the others to advantage. Get a pure white kitten, a pure black one, a red tabby, and a silver ditto. Take great care in the training of them, be careful in feeding and housing them, and you will have your reward.

The Spotted Tabby is also very pretty. He ought to be a good, sizeable animal, with broad head, short ears, and a loving face; ground colour a dark grey, one dark stripe, and down the spine, and diverging from this stripes of black broken up into spots.

Points.--The Silver Tabby ought to be--

1. *In size*, less than or about the size of the Red Tabby, and very quick and graceful.

2. *Head*: Large and shapely, but not so blunt as the Brown Tabby's; ears short and eyes light.

3. *Colour and markings*: Of a deep Aberdeen granite, grey in the ground-work, and the markings very dark and beautifully arranged. Don't forget the Mayor's Chains.

4. *Pelage*: Longish, if anything; but not so long as to make the judge suspect crossing with the Persian.

Points of Silver and Spotted Tabbies.

Size, 10.

Head, 5.

Colour and markings, 30.

Pelage, 5.

Total, 50.

There are one or two fancy cats I have not mentioned, as the Red-and-white, etc; but I believe I have said enough to make anyone, with a little study and attention, a good judge of the points and qualities of the different breeds of the English domestic cat.

*****

CHAPTER SEVEN.

Asiatic Cats.

When I was a little boy at school, floundering through Herodotus, and getting double doses of fum-fum daily for my Anabasis--for my old teacher, when he couldn't get enough Greek into one end of me, took jolly good care to put it in at the other--there was no man I had greater respect for than Alexander the Great, owing to his having done that Gordian knot business so neatly. I practised afterwards on the dominie's tawse (i.e., the fum-fum strap); I tied a splendid knot on it, and then cut it through with a jack-knife; but, woe's me! the plaguy dominie caught me in the very act, and--and I had to take my meals standing for a week.

But ever since then I have always been a don at knots; and I give myself no small credit, whether you do or not, reader, for the dexterous manner in which I have polished off the cat-classification knot. There it lay before me, interminable, intricate, incensing; and bother the end could I see to it at all at all. "Draw the sword of Scotland." Swish! There it lies, the short-haired European pussies on the one hand, and the Asiatic or long-haired on the other.

Among these latter you will find exactly the same colours, and the same variety of markings, as among the European cats proper. We give their points in a general way.

1. *Size*: The blue cats and the pure white are usually of the smallest dimensions; next comes the black, and lastly the tabbies. Some of these latter grow to immense sizes, and are animals of a beauty which is at times magnificent. The cat that belonged to Troppman, the distinguished French murderer, and now, or lately, possessed by Mr Hincks, of Birmingham, is worth going a day's journey to behold. Yet, although very large, they are very graceful, too, and can spring enormous distances. Fierce enough, too, they can be when there is any occasion, especially to strangers or dogs.

2. *Head*: The heads of the white, blue, and black ought to be small, round, and sweet, the expression of the countenance being singularly kind and loving. The heads of the tabbies ought to be broad and large, and not snouty. The whiskers of both ought to be very long, and of a colour to match the general tone. The ears have this peculiarity--they are slightly bent downwards and forwards, which gives rather a pensive character to their beauty. They are, moreover, graced by the *aural tuft*.

The eyes must also match; and this is what I like to see--a blue eye in a white Persian, a hazel in a black, and a lovely sea-green in a tabby.

3. The *Pelage*: The pelage is long (the longer the better), especially around the neck and a-down the sides; and a good brush, gracefully swirled and carried, is an essential point of beauty. The fur ought to be as silken as possible; this shows that the cat is not only well-bred, but well-fed and taken care of.

4. *Markings*: They ought to be as distinct as possible, as pretty as possible, and evenly laid on with reference to the two sides.

5. *Colour*: All white in the pure white, all black in the black, and so on with the other distinct colours; and for the tabbies the same rules hold good as those given for short-haired tabbies.

General rules for judging Asiatic Cats.--First scan your cats, remembering the difference in size you are to expect in tabbies from the others. Next see to the length and texture of the pelage--its glossiness, and its freedom from cinder-holes, or the reverse. Then note the colour, and the evenness or unevenness of the markings. The head most be carefully noted, as to its size and shape, the colour of the eyes and nose, ditto the whiskers; mark, too, the *lay* of the ear, and its *aural tuft*. In the tabbies the *Mayor's Chain* should swirl around the chest. Lastly, take a glance at the expression of face.

Merits of the Asiatic Cats.--I think every cat-fancier will bear me out in saying that, although more delicate in constitution than our European short-hairs, and hardly so keen at mousing, ratting, or so fierce in fighting larger game, there can be no doubt of it they make far nicer pets. They are extremely affectionate and loving in their dispositions, and so fond of other animals, such as dogs, pet rabbits, guinea-pigs, *etc.* Their love for a kind master or mistress only ends with life itself. Then they are so beautiful and so cleanly, and, if kept in a clean room, take such care of their lovely pelage, that I only wonder there are not more of them bred than there are. They are a little more expensive at first. You can seldom pick up a good kitten at a show under one pound sterling--but if you do succeed in getting one or two nice ones, I am quite certain you will never have to repent it, if you only do them ordinary justice.

It will be well to end this chapter here; but before doing so, I beg to make one or two remarks, which I feel sure will interest secretaries of coming cat-shows.

1. In all shows give the cats nice roomy pens, whether of wood or zinc.
2. Attend well to the ventilation, and more especially to disinfection.

3. Attend to the feeding, and, at a more than one-day show, cats ought to have *water* as well as milk. I think boiled lights, cut into small pieces, with a very small portion of bullock's liver and bread soaked, is the best food; but I have tried Spratt's Patent Cat Food with a great number of cats, both of my own and those of friends, and have nearly always found it agree; and at a cat-show it would, I believe, be both handy and cleanly.

4. On no account let the pussies lie on the bare wood or zinc, but provide each with a cushion of some sort, and have a small box filled with earth or sand, in each pen. *Sawdust in a cat's cage is an abomination.* It soils the fur, and gets into the food-dish, and renders pussy simply miserable.

* * * * *

CHAPTER EIGHT.

On Diet, Drink, and Housing.

"Throw physic to the dogs," said the immortal William. That was a good many years ago, and dogs then were of very little value, and little used either to physic or good treatment; but nowadays we have found out that the possession of even a cat, entails upon us the duty and responsibility of seeing she is well cared for while in health, and properly treated in sickness. I recommended small doses of quinine and steel to an unwell pussy the other day.

"Ma conscience!" cried her owner; "gie medicine to a cat! Wha ever heard o' the like?"

I'm sorry that woman was Scotch, but glad to say I reasoned even her round, and her cat is now as sleek and lively as the day is long.

Most, if not all the diseases which feline flesh is heir to, are brought on by bad feeding, starvation, or exposure to the weather, especially the cruel custom many people have of leaving their poor cats out all night, to seek for food and shelter for themselves. These are the cats who make night hideous with their howling, who tear up beautiful flowerbeds, rob pigeon-lofts, murder valuable rabbits, and, in a general way, do all they can to bring into disrepute the whole feline race. I declare to you honestly, there is as much difference between one of these night-prowlers and a well-cared-for cat, as there is between one of the lean and mangy curs who do scavengers' duty in Cairo, and a champion Scottish Collie.

Some men will tell you that it is unmanly to love or care for a cat; just as if it *could* be unmanly to love anything that God made and gifted with sagacity, wisdom, and undying love for all the human race! But I can point you out scores of men who are good sportsmen, fearless huntsmen, and fond of every manly sport--ay, and men, too, who are at home on the stormiest ocean, and never pale when fired upon in anger--who can both pride and prize a favourite cat. At Exeter, not long since, out of thirty-nine owners of cats, all were men except nine, and of these nine seven were married, and the two others were young ladies, while the owner of the first-prize cat *was a gallant soldier*. So much for the notion that only old maids care for cats.

Before going on to describe the diseases which afflict pussydom, we must give a few general instructions regarding her treatment while well.

And first, as to her food. Pussy will catch a mouse, and after playing with it for half an hour in a way which is very cruel, but no doubt makes it very tender, she will generally kill and eat it; but it by no means follows that mice are the cat's natural food. The majority of cats catch mice more for the love of sport than anything else. Nothing, therefore, is more cruel than to starve poor pussy, with the erroneous idea that it will make her a good mouser; it is just the reverse. My Phiz bids me say that mice-catching is long, weary, anxious work at the best, and she is quite certain she would die if compelled to make a living at it.

Feed your pussy well, then, if you would have her be faithful and honest, and keep your house clear of mice and rats.

I have lived a good deal in apartments in my time, and I have always avoided places where there was a lean and hungry-looking cat. It is a sure sign of irregularity and bad housekeeping.

Twice a day is often enough, but not too often, to feed your cat, and it is better to let her have her allowance put down to her at once, instead of feeding her with tid-bits. Nothing can be better for pussy's breakfast than oatmeal porridge and sweet milk. *Entre nous*, reader, nothing could be better for your own breakfast. Oatmeal is the food of both mind and matter, the food of the hero and the poet; it was the food of Wallace, Bruce, and Walter Scott, and has been the food of brave men and good since their day.

"Oh! were I able to rehearse
Scotch oatmeal's praise in proper verse,
I'd blaw it oot as loud and fierce,
As piper's drones could blaw, man."

But I cannot wonder for a single moment at this favourite Scottish food being in disrepute in England, because hardly anyone knows how to make it. Our cook at sea once undertook to supply our mess with a daily matutinal meal of porridge, and of oatcakes too. He was sure he could make them, because his "father had once lived in Scotland." Nevertheless, I gave him some additional information, and we, the Scottish officers, of whom there were two or three besides myself, were in high glee, and took an extra turn on deck the first morning, to give us a good appetite for the great coming double event. Then down we bolted to our porridge. Porridge! save the name, such a slimy, thin, disgusting mess you never saw! Well might our chief engineer call out:

"Tak' it awa', steward, tak' it awa'; it would scunner (sicken) the de'il himsel'!"

"But, hurrah!" I cried, "there's the oatcakes to come. Steward, where are the oatcakes?"

The steward lifted the cover from the dish on which was wont to repose our delicious "'spatch cock," or savoury curry, and there, lo and behold! half-a-dozen things of the shape and thickness of a ship's biscuit, black, and wet, and steaming, and we were supposed to eat them *with a knife and fork*! Meanwhile the ham and eggs were fast disappearing among the Englishmen at the other end of the table, and we poor Scots had to go without our breakfast, and get laughed at into the bargain.

But here, now, I'll tell you what I'll do for you, as Cheap Jack says--I'll give you a receipt by which you shall live a hundred years, and begin your second century a deal stronger than you began your first. Buy your meal from the meal-shop--no, not the chemist, my dear--taste it to make sure it has no "nip;" see, also, that it is fresh, and not ground before Culloden, and buy it neither too fine nor too round, but just a *happy medium*. Having thus caught your hare, so to speak, go home with it, and put a saucepan on a clear fire, with a pint of beautiful spring-water, into which throw a teaspoonful, or more, of salt, and a dessert spoonful of oatmeal. This is essential. Then sit down and read till the water boils. Now take your "spurckle" or "whurtle" in your right hand--I don't know the English of "spurckle" or "whurtle," but it is a round piece of wood, rather thicker than your thumb and not so long as your arm, and you never see it silver-mounted--and commence operations. You stir in the meal very gradually, to prevent its getting knotted, and you occasionally pause to let it boil a moment, and you continue this until the porridge is quite thick, and the bubbles rise into small mountains ere they escape, with a sound between a "whitch" and a "whirr," which is in itself a pleasure to listen to. And now it is ready, and you have only to pour it into a large soup-plate, sprinkle a little dry oatmeal over the top of it, and set it aside until reasonably cold. You eat it with a spoon--not a fork--and with nice sweet milk. "A dish fit for a king," you say; "A dish fit for the gods!" I resound. Now, having told you all this, I feel I have well deserved of my country; and I'm not above accepting--a hamper at any time.

Bread-and-milk, soaked, is the next best thing for pussy; and at dinner you must let her have a wee bit of meat. Lights, boiled and cut in pieces, are best, but horseflesh isn't bad; but you mustn't give her too much of either, or you will induce diarrhoea. Give her fish, occasionally, as a treat. If pussy is a show cat, a little morsel of butter, given every day, after dinner, will make her dress her jacket with surprising regularity.

Now, as to what she drinks, a well-bred cat is always particular, and at times even fastidious; but two things they must have--water and milk. They will often prefer the former to the latter. *But do keep their dishes clean.* Disease is often brought on from neglect of this precaution. Cats will drink tea or beer, and I have seen a Tom get as drunk as a duke on oatmeal and whisky. An old lady, an acquaintance of mine, has a fine red-and-white Tom, and whenever he is ailing she gives him "just a leetle drop o' brandy, sir." Tom, I think, must have had two little drops o' brandy yesterday, when he rode my fox-terrier, Princie, all round the paddock. Those naughty drops o' brandy!

Just one word about housing. There is no more objectionable practice than that of turning your cat out of doors at night, and none more certain to engender disease and spoil your pussy's morals. If you have taken the least pains to train your cat to habits of cleanliness, she will never misbehave herself. Keep her in at night, then, and you'll have her in health; keep her in if you want to run no risk of getting her poisoned; keep her in, and the neighbours will bless you. Don't lock her into a room, though, unless she has an attic to herself. Let her have the run of the house from basement to roof. Give pussy a bed to lie on, or let her find one for herself, which she has a happy knack of doing, as I daresay more than one of my readers can testify. My pretty Phiz needn't have kittened in my cocked hat, nevertheless.

So much, then, for the prevention of disease. We will now come to diseases themselves. But just let me impress upon your mind, reader, this fact--that attention to your pussy's housing, drink, and the cleanliness and regularity of her diet, will almost certainly prevent her from getting sick.

*****

CHAPTER NINE.

The Diseases of Cats.

Before describing the management and treatment of feline ailments, I may as well mention that there are three different plans usually adopted for giving a cat medicine. Pussy must first and foremost be caught--not always an easy job, as the little creature is fond of hiding away when ill. Take her on your knee, and, as you gently soothe her, envelope her, all save the head, in a woollen shawl, and then place her in some one else's arms to hold. Now, if it is a pill or small bolus it must be dipped in oil, and placed well down behind the tongue, and towards the roof of the mouth; if it is a powder, it may simply be placed on the tongue; but the better plan is to mix it first with a little treacle or glycerine; thirdly, if it is a fluid, the mouth must be held well open, and the medicine poured down the throat out of a small phial, but only a few drops at a time.

If your cat is suffering from any severe illness, such as bronchitis, and you value her, set aside a garret or lumber-room for her accommodation, for quiet is essential to her recovery. Arrange her bed as common sense tells you will best suit her comfort; don't forget to let her have plenty of clean water to drink, and a large box of garden mould in the far corner of the room. There is only one other little matter, which must not be overlooked--and, with this, pussy's little hospital is complete--Grass.

Grass.--This is the natural medicine of both cat and dog. In large doses, it acts as an emetic; in smaller, as a purgative; its mode of action being similar in both cases, namely, mechanical irritation of the muscular and mucous coats of the alimentary canal; this causing spasmodic contraction of the stomach, or increasing the peristaltic motions of bowel. Grass also possesses valuable antiscorbutic properties, and the cat, either in sickness or health, should never want a supply of it.

If pussy has been out all night at a feline entertainment on the tiles, and the excitement has produced constipation, her remedy is grass. If she has made too free in the aviary, and the feathers of the Norwich cock lie unpleasantly on her stomach, grass is her cure; or if she, at any time, feels hot or feverish, out into the garden she goes, and a little grass, taken at intervals, soon makes her feel as fresh as the lark.

Don't let your cat want grass, then; if you live in a town, and she has some difficulty in getting it, either procure it for her yourself, or, what is better, get a boxful of earth, and sow it, and call it pussy's garden. Now for pussy's ailments.

Mange.--All skin diseases in the cat, whether pustular, papular, or squamous, may be, for convenience' sake, called *mange*. Cats are very subject to skin diseases, especially long-haired ones, and those who have been the subjects of bad or careless treatment; for they are always brought about by poverty of the blood, from under-feeding, or surfeit from over-eating on dainties. Now I must warn the cat-fancier that there is no *specific* for the cure of mange in the cat, and that the cure will take weeks, and at times even months; he must therefore make up his mind either to destroy the cat at once, or set about curing her in earnest. Attend, in the first place, to her diet. It must be nourishing, but not heating; plenty of good milk, and no meat, unless she be very thin, when raw meat in small quantities may be given twice a day. Dress the skin with carbolic oil, washing her carefully next day; then try equal parts of sulphur-ointment and green iodide of mercury ointment, mixed with an equal bulk of lard. Give her arsenic internally--one drop of the *Liquor arsenicalis* twice a day, in milk, for a week, then thrice a day for another week, when you must omit it for a day or two, and then begin again. At the same time give her, once or twice a week, a little sulphur. Placing brimstone-roll in a cat's drinking-water is all a mistake, and does no good at all. Sometimes the disease will only yield to a course of iodide of potash. Give her half-grain or whole-grain doses, made into little boluses with breadcrumbs--which any chemist can make for you--twice a day.

Ulcers.--Cats are liable to a variety of these, but they can best and most conveniently be described as of two sorts--*constitutional* and *accidental*. The first are the most difficult to cure, and are usually found on the toes or feet. Confine the cat to the house for a term; any simple ointment, such as that of zinc, will do for a dressing, as it will not hurt her if she licks it. Put her on a course of arsenic, as recommended above; give her, once a week, one grain of calomel, or two or three grains of grey powder and a little sulphur; and, if the sores appear sluggish, touch them once a day with blue-stone or nitrate of silver. Feed her well and regularly.

Accidental ulcers are generally the result of scratches and wounds received in the hunting-field, or during some slight difference of opinion with the pussy over the way. They require no internal treatment. If they look angry, bathe in warm water, or milk and water, and use, occasionally, a little lotion of sulphate of zinc--ten grains to four ounces of water, to which add one drachm of tincture of lavender. If the sores are sluggish, and indisposed to heal kindly, truss the cat in the shawl, and cauterise with nitrate of silver; afterwards dress with the mildest mercurial ointment.

Inflammation of the eyes is generally the result of injury or cold caught from exposure. It may be confined to one eye, or may attack both. In either case the treatment is the same. Begin by the use of a purgative--say two or three grains of compound jalap-powder mixed in glycerine, and given in the morning; give nothing but bread-and-milk to eat, and let the cat have a little sulphur mixed with butter or lard every second day. The external treatment consists in bathing frequently with warm water or weak green tea, and the following lotion, may afterwards be used with advantage: two grains of sulphate of zinc to an ounce of water, or one grain of nitrate of silver to the same quantity of *aqua pura*.

Simple Maladies.--If you are fond of your cat you will naturally easily know when she is getting out of sorts or going to be ill. When you observe, then, from her appearing dull and apathetic, refusing her food, taking to dark corners, or sleeping all day, without attempting to go out of doors; and, especially if her coat is dry; catch her at once, and give her an emetic. Try a little salt and water first, and, if that will not act, two grains of sulphate of zinc will, given in luke-warm water. Afterwards administer as much castor-oil as you would give to a baby, or two or three grains of grey powder. Such treatment, taken in time, will often have the effect of cutting short a serious illness.

Operations.--Never hesitate to open an abscess if you think, or rather, if you are about half sure, there is matter in it. Afterwards foment with warm water. Poultices are unhandy. If the cat's leg has been severely lacerated and broken in a trap, and there seems little likelihood of its being able to heal, cut it off. Do it quietly, gently, and firmly; the ragged edge of the bone may be sawn off with a table-knife made into a saw with a file. (I cut a man's finger off the other day with the same instrument. About a fortnight after, the commander, sitting at luncheon, made the innocent remark: "This knife is rather blunt, steward. I'm hanged!" he roared, immediately after, as he dashed the knife through the open port, "I'm hanged if it isn't the doctor's saw!")

Be sure to leave enough flesh to form a flap to cover the bone; stop the bleeding with the actual cautery, then sew up and dress the wound in sticking plaster; only leave room for the egress of matter. Painful operations of this sort are always better performed under chloroform.

Lay the cat on her side (rolled in the shawl) on some one else's knee, pour a little chloroform into a handkerchief, and hold it *near, not on* pussy's nose, or you will smother her. As soon as one portion of the chloroform gets evaporated supply its place with more; in from five to ten minutes pussy will be in the land of nod.

Consumption.--Consumption in the cat is curable, because it is not necessarily disease of the lungs. The term is used to denote all sorts of wasting disease in which pussy falls away in flesh, in coat, and in general health. The treatment must be careful--regulation of the diet and attention to her housing, an occasional mild purgative and dose of sulphur-butter. You may give her raw meat steeped in wine if she will take it; but remember your great sheet-anchor in the care of all these cases is *cod-liver oil*, a dessert spoonful every day, or even more. And you may supplement the treatment most advantageously by giving, twice a day, the sixth of a grain of quinine.

One word of warning to cat-fanciers before I close this chapter. *Never ask a veterinary surgeon about your cat*. Their knowledge of canine ailments is vastly behind the times; their knowledge of cat diseases is simply and literally *carte blanche*. If you want your pussy killed or tormented to death, *go to a chemist*. The chemists in this country, through their ignorance, and impudent assumption of medical knowledge, slay their thousands annually. Their ignorant patients, however, go with their eyes open, and place themselves in chemists' hands. Well, as a paternal government refuses to protect the people, let the chemists go ahead and poison away; but, if warning of mine will be heard and heeded, they shall not poison our pussies too.

* * * * *

CHAPTER TEN.

Diseases of Cats--Continued.

Probably one of the commonest and most distressing of complaints in the cat is *diarrhoea;* and what makes it all the more distressing, is the fact that, instead of receiving sympathy and good treatment in her distress, she is often harshly treated, kicked about, and thrust out of doors.

Diarrhoea is usually brought about by want of regular feeding, by improper food, and exposure to wet and cold. Different sorts of food will also induce it--such as rancid horseflesh, sour milk, an over-allowance of fat or liver. If taken at once, the treatment is generally very successful; if let go on too long, the cat will rapidly lose flesh; and the advent of dysentery will make it a charity to put her out of the way.

Give her at first a small teaspoonful of castor-oil, to which add two drops of solution of muriate of morphia. This will often stop it, and remove all offending matter from the intestines. If there is no improvement, repeat the dose on the second morning, and give small doses of common chalk mixture three times a day, with two drops of laudanum divided between the three doses. Let her have nothing but bread and milk to eat, or a little corn-flour, if she will take it; if not, give her fish--she won't refuse that.

A few drops of solution of lime added to her milk will do good.

If she be very much reduced in weight, and has no appetite, try two grains of quinine made into twelve pills with breadcrumb: dose, one three times a day. Or you may give cod-liver oil.

Dysentery is a frequent sequel to badly-treated diarrhoea. It is simply ulceration of the coats of the bowels, combined with great emaciation, roughness of coat, dejected look, and loss of appetite. Unless a very valuable cat, I would not advise you to keep her alive. You may, however, with patience, bring her round. Give her, then, a grain or two of calomel occasionally, and quinine three times a day, unless she exhibits any tendency to fits. House her well, and give her the most generous of diet--raw meat, eggs, etc, and a little port wine daily, or even a small quantity of brandy.

Gastritis, or inflammation of the stomach, is by no means rare in the cat, and is frequently the result of poison having been given with the hope of causing death. The cat simply pines, and gets thin, and refuses nearly all food, which, when she does eat, causes pain, sickness, and vomiting. The bowels, too, are often disordered. There is nothing better, in these cases, than the tris-nitrate of bismuth, from one to three grains to be placed on the tongue twice or thrice daily. You may also give occasionally a grain or two of calomel with a little rhubarb powder.

If there is much emaciation, cod-liver oil may be tried, and a small allowance of raw meat, cut into little bits; and quinine.

Bronchitis.--This is a much more common and dangerous disease than is generally supposed. It often attacks cats at a particular age--say, six or eight months--and, indeed, is somewhat analogous to distemper in the dog. It is ushered in by the usual symptoms of a bad cold--staring coat, watery eyes, and a slight cough. If the disease be confined to the lining membranes of the nose and throat, there will be but little cough, but it usually attacks the bronchi (windpipes) themselves. There is pain, a slight swelling of the nose, and mattery exudation from both nose and eyes. After a few days of the acute comes *the chronic stage*. Pussy is now a very wretched and unhappy little object indeed. She wanders about the house coughing continually, with her little tongue protruding. She gets rapidly thin, and refuses all food; and, if not attended to, generally seeks some quiet, dark corner in which to die.

Treatment.--Great good can be done in the first stage by hot fomentations applied across the face. These must be frequent, or they are of no avail. Keep pussy indoors, and at first let her diet be low--simply bread and milk, and occasionally fish. Give her castor-oil alone, if there is no diarrhoea; if there is, add to the dose two drops of solution of muriate of morphia.

As the disease gets chronic, and pussy begins to lose flesh, do everything you can to support her strength by beef-tea, nourishing food, and wine. If the cough is troublesome, get her the following, compounded by your own chemist:--R. Extr. conii, Pil. scillae, co. aa., gr. xv.; Camph., gr. xx. Mix and make into twenty-four pills, and give one night and morning.

Latterly give cod-liver oil to complete the cure, which, in this case, will act like magic.

If the mange is present in any shape, it must be carefully seen to as directed under that heading.

Fits.--These are by no means uncommon among our domestic cats. They are of various kinds--fainting fits, delirious fits, and convulsive fits.

The former are usually caused by weakness, exposure to the weather, and general ill-treatment, or loss of blood. All that is required during the fit is rest and exposure to a current of cool air. After the fit you ought to set about getting pussy's bodily health into better condition by good food, tonics, and oil.

Delirious fits are those in which the poor cat, through mental or bodily suffering, apparently goes wild, dashing madly through the house, springing through a window, and finally hiding herself away in some dark corner. You must catch her and put her into a quiet room, and do all you can to soothe her. Apply smelling-salts to the nostrils, and bleed. This operation is easily performed by making a puncture through any of the small veins inside the ear, and fomenting in hot water. An emetic--if the cat is not insensible--will, in all probability, do good, as, both in the delirious and convulsive fits, the stomach and bowels are generally out of order.

Convulsive Fits.--The cat emits a cry as of pain and terror, and falls down on her side, foaming at the mouth, and with convulsive motions of all the limbs, accompanied with cries and moans. Usually ends in a delirious fit. During the fit do nothing at all, except prevent pussy from injuring herself or any one else; and do this gently and firmly. A pinch of snuff or smelling-bottle applied to the nose can do no harm. Afterwards bleed, and keep her in a quiet, cool room, and treat as for the delirious fit above described. When pussy has recovered--and especially if she has had a succession of fits--something ought to be done to prevent their recurrence. If too fat, you must reduce her by lowering her diet, and giving a little sheep's liver and milt two or three times a week. If too thin, tonics and raw meat must be given, and cod-liver oil every morning. If, in spite of this, the fits recur, you must have recourse to such an alterative as the following, which has done good in many such cases:--R Bromid. potass., gr. xv.; Iod. potass., Zinci sulph., aa., gr. v. Mix with moist breadcrumb, and make twenty boluses, of which the dose is one night and morning.

Jaundice.--Called also the yellows. The disease can hardly be mistaken. It is characterised by general feverishness, loss of appetite, a disposition to "lie about," and by vomiting of a bright yellow or green fluid, covered with froth.

The skin, eyes, and lips are also tinged with yellow. It is often fatal if not attended to in time.

I give, to begin with, a very small teaspoonful of Glauber salts, diluted with plenty of water. It acts as a purgative or emetic, I don't care which. If the vomiting continues, try a few grains of white bismuth placed on the tongue, or take three drops of creosote, and five of aromatic powder, and form into ten pills, with breadcrumb. *Dose,* one three times a day.

For four or five nights running give one grain of calomel on the tongue. But watch the symptoms, and omit for a night or two, if it causes too much purging. If not, you can give a small dose of castor-oil in the morning.

As she gets well, strengthen her, and encourage her appetite with quinine first--no wine--and, after a week or two, with raw meat and cod-liver oil.

Milk Fever.--Only cat-fanciers will believe that poor pussy suffers, at times, the most cruel tortures, from the thoughtless practice of depriving her of her kittens all at once. Either this or cold usually produces milk fever. I need not describe it; it being synchronous with the suckling season will be sufficient to enable even a tyro to diagnose it. If the cat is very much excited, and partially or wholly delirious, bleeding must be resorted to, and afterwards give a castor-oil purgative, with three or four drops of the compound tincture of camphor, and keep her in a quiet room. At the same time, the swollen and painful teats must be frequently fomented with warm water.

Never take a cat's kittens away all at once, but always leave one at least. If she has five, and you mean to drown four, drown two one day and two the next, so that the first milk may be well drawn off.

I have not mentioned half the ills that feline flesh is heir to, but I think I have said sufficient to indicate the *general plan of treatment* of cat diseases. Let me only just repeat that if you use your pussy well in the matter of housing, food, and drink--bar accidents--you will never have her ill at all.

* * * * *

CHAPTER ELEVEN.

Tricks and Training.

Before going on to speak of the training of youthful pussy, there is one subject which deserves a word or two at least--namely, the humane destruction of cats, when such destruction becomes necessary.

Kittens, at least, people have often to get rid of, or the whole world would be peopled with cats, and that would hardly do. Although I am no advocate for the rash and hasty condemnation of the sickest cat that ever is, still, I must confess that, at times, to destroy a cat is to be merciful to it.

Never give kittens poison, it is cruel in the extreme; you might chloroform them to death, but one doesn't like to waste much time in taking life, if merely a kitten's; the pail is always handy, and the poor wee things don't really suffer much if you do it properly. Always sink them, and keep the pail for three hours, after which bury them at once. I'll give you an example of the wrong way of doing things. Miss M--n, who lived not a stone's throw from where I now write, and who is an old maid (and may a merciful Providence keep her so!), was changing her residence last month, and at the last moment thought she couldn't be bothered with more than one of her kittens--little Persian beauties, whom she had let live a whole month--so one was snatched from its mother's arms, and pitched carelessly into a pail of water. She never heeded its cries, nor the mother's piteous appeal to save her offspring; so presently kitty was dead, to all appearance, and the bucket was emptied over the wall into an adjoining field. This was at eleven o'clock in the morning, and late that evening some boys, in passing, were attracted to the spot by plaintive mews, and there they found the kitten crawling in the grass, with sadly swollen body and inflamed mouth. The boys drowned and buried it, being more humane than old maid M--n.

If necessity, then, compels you to part by death with an old cat, and probably an old friend and favourite, I do not advise you to have her drowned. It is cruel in many ways; there is the catching of her, the putting of her into the sack with the stone, and the march to the waterside, the cat knowing all the while what is to happen, and that her mistress ordered her death. Do not drown her. If there is any one you can really trust, that you are sure knows the difference between a gun and a washing-stick, by all means have her shot. It is over in a moment.

The next best plan is to administer morphia. Don't grudge her a good dose--five or even ten grains. Cats are wonderfully tenacious of life, but they can't stand that. Make the morphia into a pill, with a little of the extract of liquorice, and force it down the throat. Pussy will sleep the sleep that knows no waking, and you will have the satisfaction of knowing she did not suffer.

Apart from teaching a cat tricks, which tend to amuse children or older folks, there is a training which every pussy needs when young--viz, to be cleanly and honest. For some weeks after the kitten has been taken from its mother, and gone to its new abode, a flower-pot saucer filled with sand, or, what is better, a small box of garden mould, must be placed in a particular corner of the room, and the kitten taught to go there; two or three lessons are usually sufficient. By degrees wean her from the box, and teach her to go out of doors.

As to teaching her the difference between *meum* and *tuum,* I maintain, with all cat-fanciers, that cats are honest by nature, although they may, at times, be tempted to steal a herring, or take a slight liberty with the canary. The great secret is to feed pussy well, and be kind to her; you may then let her sit on the table, or even extend to her the liberty of the press. Depend upon it she will never do anything to deserve disfranchisement.

If ever you catch pussy tripping, chastise her; but don't forget this, you must do so only very moderately, or in the fright she will forget what she is being whipped for. A little bit of whalebone is the best thing to use, but take care you do not hit her about the head. I have often known cats severely chastised for what they were quite innocent of. One pussy, I remember, used to be thrashed every day for a whole week for a certain act of impropriety, and it turned out, after all, that Charley, the black-and-tan, was the real culprit. She took it out of Charley, however. She whipped him upstairs, and she whipped him down, and finally she whipped him over the window, which was two storeys high. Poor Charley was much hurt, and didn't turn up again for a fortnight.

Would you have your cat a good mouser? Then *feed her regularly* and liberally; I assure you, madam, that is the whole secret.

Cats, when young, can be taught a whole host of amusing tricks.

The most graceful of these is, perhaps, leaping heights. A cat that has had constant exercise at this sort of thing will spring almost incredible distances. The best plan to train her to this is to attach a hare's foot to the end of a rod and set it in motion for her. You can every day place it a little higher, and she will soon take to it naturally. Cats thus trained will climb the tallest trees, and leap from branch to branch like squirrels.

By holding your arms in front of pussy you will soon teach her to leap backwards and forwards over them. As she gets older, increase the distance of your arms from the ground, until at last you place them right over your head, and pussy will go over and through like any old steeple-chaser.

You may teach her to go through a hoop, or hoops, held at any elevation, and in all conceivable positions. Remember always to speak kindly to her when teaching her anything. Never chastise her; and when she has performed her little feat to your satisfaction, make much of her, and give her a morsel of fish, or any favourite food.

Cats are easily taught to fish in this manner: take them when young to a shallow stream, on a clear day, where the minnows are plentiful, and throw in a dead one or two, and encourage the cat to catch them. She will soon be after the living ones.

I had a cat that I taught to retrieve like a dog, and to fetch and carry. The same cat had for its constant companion my cheeky little starling, who used to hop about and on her, pick her teeth, and open her claws, but she never attempted to molest him.

You can teach your cat to follow you like a dog, and take long walks with you, and to come to you whenever you call her by whistling.

I have told you how to make your cat a good mouser, now I'll give you another wrinkle--how to make her a good trickster--*love her* and take an interest in all her little performances, and you will be surprised at the amount of tricks she will learn.

Without reference to the accomplishments of performing cats, who require a special education, I may here enumerate just a few of the many simple performances, which, with firmness, gentleness, and patience, you may easily teach any cat of ordinary brain calibre. A cat may be taught to beg like a dog; to embrace you; to pat your nose or your neighbour's nose when told--(N.B. It's just as well it should *always* be your neighbour's nose)--to down charge; to watch by a mouse's hole; to stand in a corner on her hind legs; to move rhythmically to music; to mew when told; to shut her eyes when told; to leap six or eight feet through a hoop or over your head; to feign sleep; to feign death; to open or shut a door; to ring the bell; to fish; to swim, and retrieve either in the water or on the land.

I have a cat who, if I hold her up in front of the map of London, will place her paw upon any principal building I like to name. The cat has been used to be carried round the room to catch flies on the wall. The principal buildings in the map are marked with square black spots, which she naturally mistakes for flies, so you have only to hold her in front of the map nearest to the spot you want her to touch, and slightly elevate your voice when you name the place, and the thing is done.

* * * * *

CHAPTER TWELVE.

Agremens of Cat Life.

Before we can thoroughly understand the ways and habits of any animal, we must try, in a manner, to put ourselves in that animal's place, and thus be able to study life from its point of view.

I don't believe that God made any creature to be otherwise than happy, and He has endowed each member of His creation with just that amount of reason and instinct which shall enable it to find its food and a place to rest in, make love in its own way, marry after its own fashion--by civil contract--bring up its young, and, in a word, be generally jolly. I found a poor bee this morning getting drowned in the water-butt. "Yes," I said, "I'll save your life, but I will give you as a treat to my pet spider." Man has the proposing, but not the disposing. I laid my bee for one moment on the edge of the butt to dry, when whirr! away he darted through the bright morning sunshine, and my spider had to be content with a bluebottle for breakfast. This spider, I may tell you, is a very large and beautiful specimen, striped and marked like a silver tabby. He lives in an outhouse, and has a web, the network of which is a yard in diameter, with goodness knows how many feet of tack, and sheet, and stay, and guy. And a very amusing rascal he is, and not a bit afraid of me. Nearly every day, I give him a bee with the sting out. (It is in the kaleidoscope of events; that some day I may leave the sting in, just to see how he feels it.) I place the bee in the web, and it is amusing to see how quickly my friend shins up the rigging--he catches the bee by the shoulders, and makes him spin for a few seconds like a top, till he is completely enveloped in a gauzy shroud, and there is a big hole in the web. I tell my spider he shouldn't make a hole in the web. "Never mind that," he replies, "soon make that all right," and sure enough next morning the web is nicely repaired, and the bee nearly eaten. I don't think he eats all the bee himself. I am convinced that he has a little wife who lives somewhere in a corner, and that every day he is careful to send her a leg, or a wing, or a bit of the breast. Well, he is happy, I know. Hadn't he a nice private house, without rent or taxes, maybe a wife, and a thriving business, to say nothing at all about the bee. I have studied cats as I have studied that spider. I have imagined myself that spider. I have been, or imagined myself to be, a cat--a Tom, you know, and I can fully understand a pussy's life and a pussy's joys and sorrows.

"How different," I thought, as I mused one morning under a tree, "is the life of a cat from that of a dog. I'm the parson's cat to be sure, but then I'm my own master. Now, there is the parson's Saint Bernard dog, Dumpling for instance,--an honest, contented fellow enough, but, bless you, he isn't free. *I* am. Dumpling can't do as he pleases. I can. I can go to bed when I like, rise when I like, and eat and drink, when, where, or what I choose. Dumpling *can't*. Really I feel I can forgive Dumpling for chasing me into the apple-tree last Sunday when I think of the dull life the dog leads, and how few are his joys compared to mine. Poor Dumpling needs servants to wait upon him, and he can't even walk a couple of miles, and make sure of his way home, or sure of not getting into a row, or not getting stolen, or something else equally ridiculous. The other day Dumpling actually sat on the door-step for two hours in the rain, till his great shaggy coat was wet through and through, because, forsooth, he didn't know how to get the door opened. Would I have done that? No. I should have walked up politely to the first kind-faced passenger, and asked that passenger to 'be good enough to ring this bell for me, please, 'cause I ain't big enough,' and the thing would have been done. Could Dumpling unlatch a door or catch a mouse? Could he climb a tree and rob a sparrow's nest? or could he find his way home over the tiles on a dark night? I would laugh to see him try.

"Now here am I on this bright, beautiful summer morning, as fresh as a daisy, as happy as a king. Catch me sleeping in the house on a summer's night!

"How sweetly the birds are singing, but how much more sweetly they will taste! What a glorious day I had of it yesterday all through! Put in an appearance at the parson's breakfast-table, just for fashion's sake, and pretended to drink the milk my kind mistress placed before me. Fairly won the old lady's heart by rubbing my head affectionately against the canary's cage. 'Dear Tom,' said she, '*you* would never touch the pretty bird?' Oh! wouldn't I, though?

"What a nasty old man that Farmer Trump is! I'm sure, if it wasn't that I have a taste for pigeons, and am a little bit of a Columbarian, I would never have thought of looking at his lot, anyhow. Besides, I had only eaten two when in came *he*, and out went *I*. Well, if he didn't take his gun and fire after me. Well, if he hadn't done anything of the sort, he wouldn't have shot his bantam cock.

"I didn't go into that milk cellar of my own free will. It was purely accidental. I was chased by a dog, but being in, how could I, being only a thirsty cat, and amid such profusion, help helping myself to a drop of cream? And if the clumsy old dairymaid hadn't thrown her shoe at me, she wouldn't have broken the milk-house window. It was no business of mine. I met Master Black-and-tan outside, and warmed him. I gave *him* sore eyes. That old shoe brought luck with it, however, for about an hour after I found myself in a large and beautiful garden, filled with beds of the rarest flowers. It isn't always you get a bed made for you, thinks I; so I scraped about me a bit, and went off to sleep in the sun. Where did that half-brick come from? I wonder. I'm somehow of opinion that it was meant for me. However, if people will use profane language, and heave bricks at the heads of unoffending cats, they mustn't be astonished if they do smash the cucumber frame.

"I find it so much better to live in the free forest, because, if I live in the house, a day never passes that I do not get into a row, and I always get the worst of it. Only yesterday I looked in for a few minutes at tea-time, and there was Dumpling standing, with a yard of tongue hanging from one side of his mouth; and Master must pat him, and call him a fine fellow; then I jumped on the sofa-stool, and smacked him in the face, and Dumpling knocked down the stool to get at me, besides a cup and saucer, with his wisp of a tail, and I bolted through a pane of glass, and got blamed for that. Day before, a mouse was pleased to get behind a china vase, and I had to break the vase to get at it--I got blamed for that. Same day I ran away with a mackerel. That mackerel seemed positively to say, 'Oh, pussy, do run away with me, and eat me in some nice, quiet corner.' And I did; and, would you believe it, I was even blamed for that!

"I'm going to see Zelina to-night. Zelina is a beautiful black Persian angel, with hazel eyes and flowing fur, and a voice that would lure the larks from the sky. Zelina belongs to the barber, and I met her by appointment in the back garden, and found her very thick with three other fellows. That's the worst of Zelina. But I fellowed them! For five minutes you wouldn't have seen either of us for fluff, and at the end of that time little remained of the other cats save the teeth. Meanwhile Zelina looked calmly on. Then I wooed Zelina beneath the moon, and thrashed her, and beat her, and bit her, till at last she consented to fly with me to a foreign shore; but we made such a row that we awoke the brute of a barber, and he threw a basin of dirty water right over us, and there was no more foreign shore thought of. But I'll see her to-night, sweet Zelina!"

I'll conclude this paper with a rather curious anecdote, told me by Captain A. Brown, late of Arbroath, now of Chatham, Canada. "We have a cat," says Captain Brown, "who brought up a kitten in a loft above the woodshed, until it was old enough to wean; she then brought it down to run about, but the dog (a puppy) would on every opportunity take the kitten in its mouth and drag it about. This the cat didn't seem to like, so one day she took it in her mouth, and carried it along, on the top of the fence, to the nearest farm, a quarter of a mile off, where the *kitten's father lived*. She placed the kitten at the male parent's feet, gave it suck once more, then started off home along the fence, and never went near it again."

This anecdote, for the truth of which the captain vouches, clearly proves that pussy has a much larger amount of reasoning power than most people give her credit for. It was just as though pussy had addressed the male cat thus:

"I've brought you your youngster, Thomas. It cannot live at home for the mischievous puppy. Goodness knows I've done *my* duty to him as a mother; now, hub, you have a turn. Time about's fair-play, Thomas; good-bye."

* * * * *

CHAPTER THIRTEEN.

Sagacity of the Cat.

"The dignity of life is not impaired
By aught which innocently satisfies
The humbler cravings of the heart; and he
Is still a happier man, who, for the heights
Of speculation not unfit, descends,
And such benign affections cultivates,
Among the inferior kinds."

Wordsworth.

I think many of the miseries which the "harmless necessary cat" has to endure in this wicked world of hers and ours would be mitigated if not entirely removed, were we only to take the trouble to study and consider what a wonderfully reasoning and sensible little thing she is. "Leave the study to old maids," I think I hear some manly (?) reader exclaim. But why to old maids? It is you who are unkind to pussy, and regardless of her comforts, and not old maids. And indeed, indeed now, I never for the life of me could see why any stigma should attach itself to an old maid any more than to a cat. Most of the old maids I have known were very agreeable persons indeed, and I've spent many a quiet and enjoyable hour with old maids over a cup of homely tea. My two maternal aunts are old maids, they even plead guilty to the soft impeachment, but cheerier bodies you wouldn't meet anywhere. They go three times to the kirk on a Sunday, to be sure, and wouldn't cook a meal on that sacred day for a world. But just see them on a week-day, look at their bright smiling faces--what odds if they do try to appear a few years younger?--and ah! just see them go through the intricate figures of the mazy Reel o' Tulloch, and hear them crack their thumbs, and cry "hooch!" you wouldn't say old-maidendom was so very dreary after that. It isn't always a woman's fault if she can't get married: many, whose early affections have been blighted, would not marry if they could, for haven't they got a posy somewhere, a locket with a face, a lock of hair, and a faded ribbon which erst was bonny blue--relics of lost love, around which cling sweetest memories of the past? Besides, have not unmarried ladies more opportunities to taste the sweets of doing good, and, better still, more time to cherish hopes of happiness hereafter, which are worth a world of wedded bliss?

Cats then, like old maids, are fifty times worse than they are painted, and the reason why people don't like them is because they don't understand them. I have at this moment a large and beautiful tabby, and I positively rejoice that that cat is so fierce to everyone but me, because before I got her she was subjected to the most barbarous treatment, neither fed, nor housed, nor watered, and I believe I was the first person from whom she ever got a word of kindness. No wonder that at first she did not understand my meaning. But she does now, though she never will be tame; but if I am asleep she mounts guard on the table near me, and her purring chant is speedily turned into a low, ominous growl if any one but touches the handle of the door. Does she know that I am asleep, and that one in sleep is helpless as regards defence? I'm sure she does, for--

Cats know the nature of sleep in others.--A friend of mine has a pussy, Kate to name, who has been early trained to habits of cleanliness. When Kate wishes to get out at night she goes to her master's bedside, and mews loudly and entreatingly. To see how she will behave, sometimes her master pretends to be fast asleep, and snores loudly. "Oh!" thinks puss to herself, "this will never do;" so she invariably stands upon her hind legs, and pats his face with her gloved hand. When he gets up, she trots pleasantly before him towards a little window, which he opens for her, and admits her into the garden. The same cat for many years used to seat herself regularly every night on a chest of drawers, waiting patiently till the door of the adjoining cupboard was thrown open for her: this cupboard was a very prolific hunting-ground of pussy's. When she had kittens, and they were able to eat, she used to bring all the mice to them, and present them with that fond "murring" mew which all cat lovers know so well.

Everybody knows that cats can open doors if left off the latch, and also that they soon get up to the mechanism of the old-fashioned hand-and-thumb latch; they open this by springing up, and holding on to the hand portion with one arm, while they press down the thumb portion with the other foot.

A lady friend of mine has a large Tabby Tom who can open a room door, by standing on his hind legs and turning the knob with his teeth. This is clever, but cats even know how to *fasten* doors, at least some do; and this same *lady was once in* a cupboard, when one of her pussies came and turned on the button latch of the door, and made her a prisoner for some considerable time!

In a small village which I know, there is an old woman who lives by keeping lodgers of the more humble description. As these have often to get up and be off early in the morning, the woman always gives them strict injunctions to shut the door when they go out, for fear of thieves. One morning a lodger had forgotten to obey his landlady's instructions. Pussy, however, had witnessed the infraction of the rule, and walked directly to her mistress's bedside, and began to mew most plaintively. Nor would she be content till the woman got up, when the cat led her directly to the door. Pussy wouldn't go out, but so soon as the door was shut, led the way again back to bed, *singing*. Old women's cats are nearly always wiser than others--they get more care taken with their training, and more comfort and love. They know all the ways, likes, and dislikes of a beloved mistress, and study them just as they do their own. Indeed, some of the things I have known old women's cats do are unaccountable in any other way, but the belief that they are possessed of a very high amount of intelligence and reasoning power. No wonder our ignorant ancestors believed them possessed of devils.

You see it is just like this--when you once get a cat to love you, you, and you only, will become the study of her whole life. She soon finds out what pleases you, and what vexes you, and also what you love, and, whether that be dog or child, she will love it too, to please you.

Cats will often, very often--just like dogs--lead those they love to places where something or some creature is in danger. It may be, as happened to myself once, while residing in Lincoln, two summers ago, when a cat came towards me out of an entry, and, as plain as any animal could speak, gazed up into my face, and cried: "Come, oh come and help me?" I followed, and she led me down the garden to a closet, through which her kitten had dropped into the cesspool below. Now just think for one moment of the amount of sagacity shown in this case! Piteously the little kit had mewed to her mother: "Mother, mother, come and help me?" Pussy's answer had been: "My dear, I can't, but I'll soon find those who will." And that was precisely my answer to the mother cat, when I saw the state of affairs, and I kept my word.

And once again a pussy--this time my own--led me a long way from my work to a distant outhouse to see her kits. After she got me to the spot where they were, she rolled on her back and held them up one by one to be admired.

I knew the case of a cat bringing her mistress hastily to a room where her sick child lay. The child had rolled on to the floor, and would have been smothered, except for pussy's timely aid.

Some will hardly credit this, because they do not see the working of the internal machine--pussy's mind--nor know the motive power--love, love, love. *Amor vincit omnia*.

* * * * *

CHAPTER FOURTEEN.

Cats Feeding the Sick.

"Ma conscience! mither, it kens its name?" Such was the exclamation of a little ragged and kilted urchin, in the remote Highlands of Argyllshire, as he heard me call my dog to give him a drink. The day was exceedingly warm, and we had had a long walk over the mountain, and had been kindly invited into a shepherd's hut, and asked to partake of a draught of cool, sweet whey--the very best of summer beverages. Nero was having a "talkee-talkee" with some rabbits, and didn't see his whey until I called his attention to it; hence the wondering urchin's exclamation.

"Hoo shouldna he?" said the mother; "poor wise-lookin' beast. Ise warrant he kens mair than that."

The idea of even a child thinking it strange Theodore Nero (the Newfoundland champion) should know his name was so amusing that I gave the boy "twa bawbees" on the spot.

And just on a par with this boy's ignorance, is the unbelieving ignorance of some people who doubt everything they cannot understand, however well authenticated. This doubting implies an assumption on their part that the knowledge they possess is the highest attainable, that their minds are, in fact, complete in themselves. It is people of this class--fools--who doubt the existence of even a Supreme Being. I read in a late number of the *Live Stock Journal* an account of a cat, which, seeing its master sick in bed, and unable to move, brought a mouse to him, and on her master pretending to eat it, the same day brought him a striped squirrel; and every day, until he got well, brought "game" of some sort and laid them on his bed.

I believe I, myself, was the first who ever *dared* to publish a case of the same kind. The story was this: A poor ploughman, who lived in a little hut at the foot of the Moffat Hills, in Scotland, fell sick of a long, lingering illness--and when the poor are ill they are poorer still; it is then the shoe pinches. This poor man had nothing in the house but meal and milk. The doctor said he must have wine. His wife pledged her marriage-gown to get it. The doctor said he must have meat. That was beyond their power to procure. But a merciful Providence had willed the man should live; and one day the little tortoiseshell cat, which was a great favourite with the poor ploughman, and had been very dull and wretched since his illness, brought in a rabbit--a thing, mind you, she had never done before--and placed it on the bed. She appeared to brighten up as she saw it skinned and cooked by the ploughman's wife, and partaken of by her sick master. And next day she brought another, and so on, almost every day, a rabbit or a bird, until her master was well, after *which she brought no more*. I took very considerable pains to test the truth of this story, and went to some expense about it as well, and found it in every whit true as first related to me. (See "Cats," by same Author. Dean and Sons, Publishers, 160a, Fleet Street.)

Since then I have had one or two cases precisely similar to the above, in which cats brought their "game-bag" to the bed of a sick master or mistress.

It is indisputable, then, that such things have been done over and over again. And now the question comes to be, how are we to account for it? In ancient times, these poor, affectionate pussies would doubtless have been condemned to death as being witches in feline form.

In our own day such cases are usually put down to a special interposition of Providence. Now, without doubting for a moment that there is a Divinity which shapes the end, we must remember that that Divinity works more by simple laws than miraculous means, and consequently endeavour to account for the occurrences in a natural way.

Cats, we know, after they have weaned their kittens, are in the habit of bringing them mice, etc, by way of food. This we do not think at all strange, and we put it down to that much-abused term--instinct. But the following anecdote shows, I think, something higher than mere instinct, and will help us to understand why the cat will bring food to a sick master or mistress.

A certain cat had kittens. They were all drowned except one, which, of course, became a great pet with pussy, who, after putting it through a course of milk, put it through a course of mice, according to the custom of country cats. The kitten grew up into a fine large Tom, and was big enough to thrash his mother, which I'm sorry to say the unfilial rascal sometimes did. But a day came when he had need of that mother's love. Tom had his leg torn off in a trap, and was confined to his pallet of straw for several weeks, and never, one single day of his illness, did his mother miss bringing her wounded son either birds or mice, until he was able to run once more, though on three legs, to go and hunt them for himself. This cat is living still, I believe. It is quite evident that a cat's affection for, and attachment to, a beloved master, are quite equal to their love for a grown-up son, and the same feelings which prompt her to minister to the latter when ill, and unable to move, would cause her to attend on the other.

Cats easily know when any one they love is sick or ailing. I returned home a few years ago, after an absence of some six months, very bad indeed. I thought I was a "gone coon," as the Yanks say, and didn't feel to have any more flesh on my ribs than there is on those telegraph wires. Well, my pet cat was rejoiced to see me, and hardly ever left my room. She would never leave me, it is true, but still there was something very strange in her behaviour. For she must have seen something strange in my appearance. Whether she took me for an impostor or not, I cannot say, but she always sat facing me whenever I was seated, seldom taking her eyes off my face, and her brows were lowered as if she were angry with me about something. What were pussy's thoughts? I asked this question one day of my father's housekeeper. "The cat kens ye'er no lang for this warld," said Eppie; "gin I were you, I'd just mak' my callin' and election sure." Calling and election! How I hated the old rook! Cats have an idea that when any one is ailing, it *must* be for want of food. Poor things! How often they suffer hunger and privations themselves, goodness only can tell! This idea is not confined to cats alone. Dogs, at least, I know possess the same notion. I could give many anecdotes to prove this, but as this book is presumably on cats, I must only give one.

An Inverness-shire student was returning from the south, and with him his faithful Scottish collie. In the Highlands there are generally two roads, the high and the low; the low road being the longest and of course the safest, and the high much shorter, but usually leading through some ugly bits of country, which are far from safe even by day, and much less by night. It was a beautiful night, quite clear and starry, with just the slightest crust of snow on the ground, barely enough to darken the heather. But such being the case, the student thought he could easily venture to cross by the hills, and thus save a mile or two. Early next morning, a woman at a neighbouring farm was surprised, while baking bannocks, by the entrance of a strange collie. The collie did not use much ceremony, but simply stole the largest bannock, and fled. This, of course, was not thought much of. The dog was hungry, and the morning cold, and he was welcome to the bannock, although it would have been more satisfactory for both sides had he asked for it. The same dog returned, however, in a few hours, and his behaviour was so strange that one of the family was induced to follow him. The dog led him a long way over the mountains, and at last brought up at the foot of a precipice, near a stream, where "something dark was lying." This something dark was no other than the poor student, who had slipped his foot on the previous night, and tumbled over the rock. He was at first supposed to be dead, but soon revived, having merely fractured a thigh, and become insensible from the cold; but the strange part of the story is to come--the bannock, all untouched, reclined against the student's cheek, *placed there by the dog*. (At page 83, volume three, "Annals of Sporting," an instance of collie-dog sagacity very similar to this is given.)

Not only do cats know sickness in others, but they are acquainted in some way with the mystery of death. Observe a cat, for instance, that has played with a mouse until she has killed it. Just see the critical way she turns it over and over with her foot, and glares into its glazing eyes. She wants to make sure the wee thing is not shamming; but, being satisfied, mark her as she coolly stretches herself, or walks slowly away from her victim, as much as to say: "Well, I've had half an hour's good fun, anyhow. Might have eaten it as long as it was alive, though; but I can't bear a dead mouse. So it's just as broad as it's long."

* * * * *

CHAPTER FIFTEEN.

Tom, Timby, and Tom Brandy.

"The best-laid schemes o' mice and men
Gang aft agley,
An' leave us nought but grief and pain
For promised joy."

Burns.

And if the schemes of mice and men often "gang agley," it is not to be wondered at that the sagacity of the domestic cat is sometimes at fault. A very large and beautiful cat, belonging to a lady in Dumbarton, was very much attached to its home--more so, perhaps, in this case, than to its mistress, for one day, much to pussy's disgust, disreputable-looking men in aprons--so pussy thought them--came to the house and began to remove the furniture. Pussy sat on the hearthrug, washing her face with a spittle and musing. "I've been so happy here," she was thinking; "I know every mouse's hole in the house, and the places in the garden where I can hide to catch the sparrows, and the gaps in the hedge through which I can bolt when that Skye-terrier chases me, and the whitethorn bush beneath whose scented boughs I meet dear Tom in the moonlight. Oh! the thoughts of leaving Tom--no, I cannot, will not, leave the old house. Missus can hang herself if she likes. Happy thought, I'll hide--hide in the linen drawer, till this cruel war is over, and then come forth, mistress of all I survey." And so she did; but, unfortunately for her calculations, the chest of drawers was moved as well; and when at last she did "come forth," much to her bewilderment she was in a house which she had never seen before in her life.

The following anecdotes may not be thought uninteresting; they are taken almost at random from hosts of others in my possession, or, if there has been any choice in the matter, they have been chosen because the three cats, whose stories here are told, lived in widely different parts of the globe, clearly proving that a cat is a cat all the world over. We'll give the English cat the preference. There is nothing very wonderful in his history. Tom was born and bred in Gloucestershire; he was presented to his master and mistress, the former of whom was a schoolmaster, when quite a little kitten, and soon became a great favourite with both. Tom, who was a tabby, soon grew in strength and beauty, until there were few male or female cats in the neighbourhood who did not own him lord and master. But Tom was so fond of his owners that he spent but little time either fighting or courting, much to his credit be it said. About this time, his master and mistress used to make frequent visits to a neighbouring village. Tom was not permitted to accompany them; but, whatever time they returned, by night or by day, wet weather or dry, poor Tom always met them nearly a mile from their own house.

Tom was remarkably fond of the schoolchildren, and every day, as regularly as the clock struck twelve, at which hour the school was released for the forenoon, Tom presented himself all ready for a romp. The family dinner-hour was one o'clock, and Tom never failed to attend. There was a knocker on the door, and whenever pussy found the door closed, he used to *jump up and knock,* just as he had seen strangers do.

Tom knew the days of the week, for he was never known to set out for school on Saturdays or Sundays, for the simple reason that he knew the school was closed.

Another strange trait in Tom's character was his fondness for poultry. "He would feed with *very young* chickens, and with the ducks and hens, never attempting to molest the weakest of them, but would even yield to them, and frequently leave the choicest bits for them." Tom's life was a very happy one until his owners removed to Leamington. Here, in the same house with him, were a parcel of rude, badly-bred children, who persistently ill-treated the poor cat, till at last Tom was missing; and it was found he had taken up his abode in a fowl-house among his old friends. This was rather a down-come for the poor cat, and he must have felt as wretched as a human being whom, after living for years in luxury, misfortune had at last condemned to the poor-house. Being removed back to his owner's house, and the children still continuing their persecutions, Tom fled to the woods and became a bandit, and no doubt met with a bandit cat's death, and died in a trap. So we leave him.

Tom Brandy was an Australian miner's cat. The miners baptised him in *aguardiente,* and hence his name. He was a beautiful large black cat, with one white spot on his chest, invaluable as a hunter, and came down like a whirlwind on every dog he saw. He was a good example of the travelling cat; he would follow his master every Sunday in Melbourne to church, hide in a neighbouring garden till the preaching was over, and then trot home behind him. He would lead like a dog in a string. Tom's travelling carriage was an old gin case. Into this Tom would jump whenever he saw preparations made for striking the tent, and lie there without ever appearing, at times for a whole day, until the new camping-ground was reached. Yes, a wild life Tom led of it in the Australian bush. When Tom's master left for "merrie England," Tom proved himself just as good a ship cat as he had been a miner's puss. Only, mind you, Tom liked his comforts when he could get them. It was no business of his if his master and family chose to be intermediate passengers. He knew better, and attached himself to the cabin, although, to show he did not forget his owners, he used to pay them a visit every evening, to see, I suppose, if they had everything they wanted. On the arrival of the ship at Birkenhead, the purser, after offering two pounds for Tom in vain, stole Tom Brandy; but Tom was at his master's house that night, nevertheless.

Tom's future home was Montrose, where he lived for two years happy enough, after which he mysteriously disappeared, and was not seen again for nineteen months. Where had he been? What had he been doing? How had he lived? *N'importe*! Tom Brandy turned up again very thin and very angry, and wanted to fight everybody save his own master. Tom lived happy ever after--that is, for three years, when he laid down upon a shelf and died like a Christian. And the days and years of Tom Brandy's life were sixteen and over, and he weighed a little under seventeen pounds.

Timby is also a Tom cat, and lives at Dunbeath Castle, Caithness; a pretty black-and-white animal, weighing about ten pounds. Timby is the coachman's cat; and as his master lives in a retired part of the country, the two are naturally very much attached to each other. Timby follows his master round the grounds and policies just like a dog. When little more than a kitten he proved himself a perfect Nimrod among cats, brought down birds from the highest trees, tore up moles from their tunnels, and was death upon rats and mice wherever he saw them.

Since he has grown up to years of discretion, Timby has learned to despise such paltry game as mice or rats. The Highlands of Scotland, as the reader doubtless knows, are infested with rabbits, and many a poor farmer is ruined by them; and these Timby makes his special quarry. It is his habit to stay out all night, and he seldom appears without a coney in the morning. If his master will accept the rabbit, Timby is very much pleased. If his master won't, and pushes it away with his foot, "Oh, very well," says Timby, "I'll have the rabbit; you have that herring of yours--I question if it will keep another day;" and he trots off with his prey.

Three years ago his master got a nice retriever dog, and to this dog Timby was at first exceedingly cruel, but latterly he grew very much attached to it; and as often as he can spare a rabbit he brings it to the dog's kennel, and seems pleased to see him devour it.

Like my own cat or cats, Timby will defend his master with his heart's blood. One day when Mr McKenzie, Timby's master, was trying a new terrier with a rabbit, Timby, who had followed unperceived, as soon as he heard the rabbit scream, doubtless came to the conclusion that his master was in danger, and sprang fiercely on another dog which Mr McKenzie was holding. The battle was short and bloody, and the poor dog had to retire very much worsted. Another day, when the coachman and his cat were lying together on the grass, a friend came up, and was just in the act of throwing himself on the turf likewise, when Timby flew upon him and lacerated his face very severely, and it was with some difficulty his master got him off.

Timby goes regularly to the sea with his master to swim the dogs, but does not himself take the water. But in coming home a rabbit is often started. Then away go the dogs, and away goes Timby, and, strange as it may seem in rabbit-coursing, Timby would gain as many, if not more, points than the terriers. However, there is no sort of spirit of rivalry betwixt them, and if the dogs choose to beat a field for rabbits, Timby stands by to catch them; again, when the dogs prefer to "lay by," Timby with pleasure goes and beats the field for them.

If Timby knows there is any vermin in a burrow, he has patience enough to wait till he secures it! and he has been known to lie near a hole *for nine hours* in a stormy day, before his patience was rewarded.

* * * * *

CHAPTER SIXTEEN.

Some Traits of Feline Character.

We all know that almost any dog that has lived a reasonable number of years, and isn't a kennel dog, but one of the family, as it were, understands pretty nearly all that is said in his presence, if it at all concerns him. My Theodore Nero is exceedingly 'cute in this respect. When I have to go out without taking him along with me, he will lie listening attentively, with just half an eye open, till he finds out in what particular direction I mean to go. After I leave home he tries every trick and wile to get round the servant, and generally succeeds; so that, on turning a corner of the road, ten to one I find the identical dog I left asleep in the parlour, coolly waiting for me. Indeed, I have often to leave my orders about him in bad French, as my wife doesn't understand good Gaelic. I get to windward of the dog that way, and, I fear, sometimes to windward of the wife too; the haziness of my French leaving the one just as wise as the other.

Till very recently, some people wouldn't even admit that a cat could know its own name; some people get wiser every day, and I, for one, believe that cats know fully as much of what we say as dogs do. As an instance of this, I give you the following anecdote, which may be entitled:

A Cat with a Conscience.--A certain Mr Coutts, of Newhills, Aberdeen, is very fond of both cats and poultry, and studies the tricks and manners of both. He recently had a hen with a large brood of chickens, the number of which day after day became lessened by one at least. The place was always searched, but not the slightest trace of a dead one could be discovered. The poor cock was blamed, ravens were suspected, and hawks deemed guilty; but still there was some mystery about it, and the chicks went on getting fewer and fewer. About this time it was observed that whenever the subject was brought up, the favourite cat seemed all at once to grow exceedingly uneasy and restless, and finally bolted off through the nearest open door. This naturally aroused suspicion. Pussy was watched, and found one day in the very act of walking away with a chicken.

I have another anecdote, something similar, of a cat called Polly. Polly had one failing, although otherwise a virtuous cat, and extremely honest--she could not resist the temptation of stealing a bit of cheese, whenever she could do so unperceived. But note the slyness of this pussy: she could never be prevailed upon to touch cheese, even if offered to her in the presence of any one of the family, evidently reasoning thus with herself: "If I pretend I can't eat cheese because it disagrees with me, they will never blame me for stealing it, and I shall often find myself locked in the same room--glorious thought!--with a whole Cheddar."

It is a well-known fact that dogs often take particular dislikes to certain people. They appear, in many cases, to be much better judges of character than we ourselves are. I believe this instinct, or whatever else it is, is not confined to dogs alone, but is equally shared by other animals. Cats, I know, possess it in a very remarkable degree. They know by some means, which I will not pretend to understand, those individuals who have a soft side towards them. Why, for instance, did that strange cat at Lincoln single me out from dozens of people who were on the street, and ask me to go to the rescue of her kitten?

Why do cats often pass other people by, and come up to me on the pavement, requesting me to ring the bell, that they may get in out of the wet? There are two strange cats who sleep in the sun almost daily in a corner of my front garden. If any one comes along they bolt at once, but when I pass up and down, they merely look at me and lie still; and I never speak to them, unless, perhaps, just a passing word. But, what is still more strange, Theodore Nero walks up and down past them without causing them the slightest alarm. Yet, what a tremendous monster he must appear to them! They just look at him, wonderingly, as much as to say: "Oh, you great, good-natured-looking brute, however you can catch mice and sparrows enough to fill your enormous stomach, I can't tell?"

I know a lady who is very fond of cats, and when out walking or shopping in town, it is quite a usual thing for her to be accosted by some poor half-starved waif or stray, and very often she goes into a shop and buys food for them, for which, no doubt, they are grateful, and for which, no doubt, she will one day receive her reward from Him who careth even for the humble sparrows. This lady was passing a house one time where a poor cat was confined, the usual occupants having gone to the seaside, and left pussy shut up in the empty house. As soon as she stopped at the door of the house, the cat's cries were quite pitiable to hear. As soon as this lady left the door, the cries ceased, only to be renewed whenever she returned. But pussy did not make the same noises when others stopped in front of the door.

A Cat deserting one Home for another.--A tortoiseshell-and-white cat, belonging now to a friend of mine, came into his possession in rather a singular way. The cat was originally the property of a neighbour of my friend, whose house was on the opposite side of the street, and about thirty yards off. There she stayed, apparently perfectly contented and happy, until she became the mother of four kittens. Then, for some reason or other known only to herself, she determined to shift her quarters, and one day my friend was astonished to see Kate, as she was called, march into his house with a kitten in her mouth, which she deposited in a safe and comfortable corner, and then set off for the others, which she brought one by one. Remember this, the cat had never been in my friend's house before! Kate's kittens were taken back again to her old home, and Kate marched them all over again to the home of her choice. And this was done every day for a whole week.

"It's no earthly use, you know," Kate seemed to say. "What I says I means, and what I does I sticks to."

And so my friend had to adopt both Kate and her family, previously having failed in an attempt to starve her out, for Kate had adopted a system of house-to-house begging, but always came home in the evening.

This cat for fourteen years used to sit patiently on the arm of her master's chair until dinner was done and she was helped.

It is exceedingly rude, I know, to doubt a lady's word, but *can you believe* what follows? 'A lady assures me that she has such an inexplicable and innate antipathy to cats, that if she enters a strange room she can tell at once if there is a cat there, whether she sees it or not. And if a cat is carried suddenly into a room where she is, she "faints dead away."

Another lady friend of mine, who is very fond of animals of all sorts, while living down in Brighton last October, was hastening home one evening just about dusk, when she suddenly found that she was not alone, but accompanied by some little black creature, which, immediately she came under the gas-lamp, she found was a poor little stray kitten. As this wee puss bounded into the house as soon as the door was opened, of course she believed it belonged to the house. Going to her bedroom to dress for dinner, there was little Miss Puss sitting on the bed singing, and apparently perfectly satisfied with her new quarters, for the lady soon found it did not belong to the house.

Pussy was treated to a saucerful of milk, and then sent adrift out into the street, chased out with a broom, in fact, for the housemaid hated cats. This kitten didn't mean to be put off like this, however. She stopped out all night, certainly, but quietly came in with the charwoman at five o'clock in the morning, and came directly to my friend's bedroom. There is no getting rid of a cat when it once concludes to board itself upon you, and this little waif soon established herself for good at Ashburnham House. But here is the strange part of the business. She seemed to know that my friend Mrs W. was only a visitor here, and constantly showed great discretion, by sticking close to her apartments and back-yard. Just once she ventured down to the kitchen, and the old residential cat bit a piece out of her ear. "If that is how you treat visitors," said kitty, "I'll stick to my own rooms in future." And so she did.

It is sometimes rather a difficult thing finding suitable apartments when you are accompanied with pets. It takes considerable tact, I can assure you, to convince Mrs 'Arris, or whatever is the name of your intended landlady, that your Newfoundland is so clean that you never can see even a hair on the carpet; that your Pomeranian is an angel in canine form; that your Persian cat wouldn't steal, if surrounded even by the most tempting viands; that your macaw doesn't scream loud enough to give all the terrace "an 'eadache;" and that your white rats never escape and run all over the house. Mrs W. had some difficulty about her kitten when she went to the lodgings she had taken at Norwood.

"I certainly did expect," her landlady observed, "a lady with birds, and a mouse, and a very large dog; but a cat I couldn't have, because I've one of my own."

Mrs W. of course promised all sorts of impossibilities regarding her pet, and her landlady finally gave in.

But, strange to say, this very house became the kitten's future home, for the landlady's grandchild struck up a friendship with the wee pussy, and when the child fell sick, the kitten would hardly ever leave her little crib, nor would the child bear Miss Brighton, as she called her feline favourite, out of her sight for a single moment. Who shall say how far the simple companionship, of this loving and affectionate wee kitten, might not have tended to the child's restoration to perfect health?

* * * * *

CHAPTER SEVENTEEN.

Love of Children and Affection for Owner.

There is hardly a domestic animal we possess that is not fond, to a greater or less extent, of children. How carefully a horse will pick his steps if a child happens to fall amongst his feet! I saw a bull one day escape, wounded and furious, from a killing-house, and dash madly along the turnpike road. He knocked down and injured several people, who could not get quickly enough out of his way; then there stood, paralysed with fear, and right in the wild brute's path, a child of tender years, which everyone who saw it gave up for lost; but the bull, who did not hesitate to attack grown-up people, suddenly veered to one side, and left this child unhurt!

My large Newfoundland dog is in the habit of careering along the street with a speed which, considering his size, is quite incompatible with the safety of the lieges. Policemen, especially, very often find themselves in the line of his rush, and Nero never hesitates to run clean through these men, so to speak, leaving them sprawling on the ground with heels in air; but the other day this dog, on suddenly rounding a corner, found himself confronted with four little toddling infants, who, hand in hand, were coming along the pavement. There was no time to slacken speed, and to proceed was certain death to one or more of the poor children, and what do you think this noble fellow did? why lifted himself clean off the pavement, and sprang high and clear over their heads.

The same dog was once in a hotel, when a friend of mine offered him a biscuit. Master Nero wasn't hungry; he would neither eat the biscuit from my friend's hand nor from my own, but when the landlord's pretty little daughter came running in, and threw her arms about his neck, and caressed him, he hadn't the heart to refuse the biscuit from *her* hands, and even accepted several from her, although still refusing them from us.

But the domestic cat is, *par excellence,* the playmate and friend of childhood. What is it, indeed, that pussy will not bear from the hands of its little child-mistress? She may pull and lug pussy about any way she pleases, or walk up and down the garden-walk with it slung over her shoulder by the tail. If such treatment does hurt the poor cat, she takes good care not to show it. It is amusing enough sometimes to watch a little girl making a baby of her favourite pussy. They are wearied with gambolling together on the flowery lawn, and playing at hide-and-seek among the shrubbery, and pussy "*must* be tired," says little Alice. Pussy enters into the joke at once, and seems positively dead beat; so the basket is brought, the little night-cap is put on, the shawl is carefully pinned around its shoulders, and this embryo mamma puts her feline baby to bed and bids it sleep. There is always two words, however, with pussy as regards the sleeping part of the contract, for little Alice never can get her baby to close more than one eye at a time. Pussy must see what is going on. Anon the baby "must be sick," and pussy forthwith appears as if she couldn't possibly survive another hour. Bread pills are manufactured, and forced over the poor cat's throat, she barely resisting. Then lullabies, low and sweet, are sung to her, which pussy enjoys immensely, and presently, joining in the song herself, goes off to sleep in earnest.

And Alice, pussy's friend, although at times she may use the furry favourite rather roughly, is kind to her in the main. Doesn't pussy get a share of Alice's porridge every morning? doesn't she sup with Alice every night? and do you think for one moment Alice would go to bed without her? Not she. And still this cat, may be as savage as a she tiger, to every one else in the house save to her little mistress. Just let you or me, reader, attempt to hold her up by the tail--well, I would a hundred times rather you should try it than I.

The very fact, I think, that faithful pussy is so fond of our innocent children, and so patient and self-denying towards them, is one reason why we should be kind to her, and study her comforts a little more than we do.

But probably one of the most endearing traits in the character of the domestic cat is her extreme attachment to, and love for, the person who owns her. If you once get your cat to really love you, no matter how fond she may be of the home where she was born and reared, she will go with you, if you but say the word, to the uttermost parts of the earth. My poor old favourite, Muffle, has travelled many, many thousands of miles with me by sea and land, and always watched over both me and my property *with all the care* and fidelity of a Highland collie. Been lost, too, she has, many a time in the midst of big bustling cities which were quite strange to her--been lost, but always turned up again.

I know of many instances in which cats have so attached themselves to their owners, that, when the latter have died, they have refused all food, and in a few days succumbed to grief, and gone, I fondly hope, to meet the loved one in a world that's free of care.

"But the largest cat," writes one of my numerous correspondents, "I ever saw belonged to my mother's mother, and was wise and sedate in proportion to its size. Its good mistress was often distressed with palpitation of the heart, and during the silent hours of night paced the bedroom floor in pain--but not alone, for the faithful creature would walk slowly at her side, seeming by his look to pity her condition, and when she lay down he would still stand sentinel at her head. He never could be persuaded to leave the house while she lived, yet a few hours before her death he suddenly took flight, but only to the lower apartments, which my parents occupied, and from which he never stirred again."

I never think, somehow, that a fireside has the same cheerful look of an evening unless there be a cat there, to sit on the footstool, and sing duets with the tea-kettle.

And I do not wonder at old women, whose friends have all long since gone before, and who have no one left to care for them, getting greatly attached to a faithful pussy; for people must have something to love.

"But, fancy loving a cat!" I think I hear some churl remark.

Yes, cynical reader, and I have, myself, before now, often shared my heart with stranger pets than cats; and I don't mind betting you that what I have left of it is bigger than yours now.

Figuratively speaking, I think a man's or a woman's heart is like a blacksmith's arm--*it grows with use.*

* * * * *

CHAPTER EIGHTEEN.

Hints upon Breeding and Rearing Cats for Exhibition, and a Word about Cat-Shows.

At nearly all the cat-shows which I have visited of late, I have been invariably impressed with this one idea: here, in these shows, we see pussy as she is in the present day--the live mouse-trap, the barn cat, at best the fireside favourite--but, at all events, the animal, of all our domestic animals, that is least cared for, and the only animal we possess, whose improvement in condition and species we have never cared to study. What this animal--the domestic cat--can become, the perfection to which she may attain through judicious selection and careful breeding, it is for future years to show.

Other nations--such as the Persians and different other Asiatics--know far more about the domestic cat than we do, and quite put us to the blush with their splendidly-bred and high-blooded animals.

It is one of the many popular fallacies current in this enlightened land of ours, that there is in the cat a certain number of bad qualities--a certain spice of the devil, so to speak--that never can be bred out. This is simply absurd, for there is no animal that lives and breathes on God's fair earth but is susceptible of improvement, both physically and morally; for, remember, a cat, little as you may think of her, has a mind *and a soul*, as well as you have. She has thought, and memory, and reasoning powers; she can love and she can fear, can be happy and gay, or sad and sorrowful, and she knows something too of the mystery of death.

With all these qualities will you tell me that she cannot be improved? I say she can; even as to race; for what can be accomplished with individual cats, may be accomplished with the whole race. I can introduce you to dozens of cat-fanciers in this country, who have made the peculiarities of pussy's nature their study, and who find that they can, at will, not only improve the physical condition of their cats; but even, by careful training, occasional gentle correction, kindness, and good-feeding, raise them from good to better, and wean them from the ways which are so objectionable in other, or merely half-domesticated cats. And, look you, the progeny of such animals--by a law well-known to all breeders--take after them, or inherit the good qualities of their parents. Hence, I repeat, if you can improve the individual cat, through time you may improve the *genus*. That time may be long in coming--

granted; but that the lovers of cats, in this country, have boldly seized the bull by the horns, and are taking a step in the right direction, is a positive fact which admits of no denial.

Now, to those who are fond of cats, and would fain improve the particular breed they have a fancy for, and probably win prizes at our great shows, I beg to offer the following hints:--

First. Having made up your mind as to what particular breed you mean to go in for, stick by that breed for a time, at least, and go in for no other.

Secondly. Be careful in your selection of parents. For instance: we will suppose you mean to breed pure white Angoras; well, purchase at a first-class show a Tom kitten and a queen kitten *from different litters*. Choose the liveliest, biggest, and most healthy-looking kitten of each litter, not, as in choosing pups, the heaviest and sleepiest-looking. The funny kitten turns out the best cat, and is more easily trained than a sulky or frightened one.

Having gotten your purchases home, remember that the royal road to a kitten's affection is straight through its stomach. Be, yourself, then, the first to present pussy with a saucer of warm, creamy milk.

Thirdly. How to get size. This is accomplished by the quantity and quality of pussy's food, and the regularity with which she gets her meals. Whatever you give a young cat, and a growing cat to eat, do not let it be too abundant. Never let her gorge herself; give her little and often. Don't let her want for a saucerful of pure water, to which she can always find access. Let her allowance of milk be put down to her and taken up again when she has had all she wants; what she leaves had better be given to the pigs. Bad milk is a fruitful source of diarrhoea, dysentery, and some forms of skin disease. A little sulphur--about as much as will lie on a fourpenny-bit--should be given at least once a fortnight, or half that quantity once a week.

Train your cats early to habits of cleanliness. Don't forget the flower-pot saucer; and remember that, if the cats you wish to take prizes with, belong to any of the finer breeds, they *must* be parlour cats, and not kitchen-bred brutes.

If you want your cats to grow large, let their food be nourishing but not stimulating; boiled cow's or sheep's lights they can eat their stomachs full of; but avoid beef, it is too gross and heating, and don't patronise the cat's-meat man.

Kittens and growing cats, in order to grow large, must have plenty of exercise and fun. Leaping exercise is best. Teach them to jump through a hoop, and keep them at it. They ought to have a ball as a toy, or a hare's foot; and ridiculous as it may seem to many, it is a positive fact, that cats--especially queen cats--thrive best who have a looking-glass conveniently placed to admire themselves in, and to wash and dress in front of.

"Ilka little maks a mickle," is a good old Scotch proverb, and believe me it is attention to little matters, to minutiae, which makes one successful in properly rearing any animal.

Fourthly. How to get Good Pelage on a Cat. The feeding of course has much to do with the length and gloss of the coat. Fish I have found is good for the coat, and a mixed diet generally, with not too much vegetables to scour them. But your sheet-anchors, after all, are the brush and the comb. The comb must be fine, and not too close in the teeth, and it should be used gently, after which brush the coat briskly all over with a long-haired soft hair-brush--a baby's brush in fact. The comb is not only a gentle stimulant to the skin, but it prevents matting, while the brush removes dust, and gives a nice glitter to the pelage. Both together act as a charm.

Fifthly. In cats other than white you will find that certain kinds of food strengthen the colours of the pelage. I am convinced, for instance, that boiled bullock's lights do, and so does sheep's blood. This fact is perhaps worth knowing. I am making experiments with other foods and some condiments, but am not yet in a position to state results.

Sixthly. Breeding for colour. No matter what colour your parent cats are, you will occasionally find waifs and strays in a litter that you will wonder to find of a different colour. But do not be discouraged; stick only to the true colours, and you will find in time that such anomalies will become few and far between. Be careful to avoid the possibility of any litter of kittens having more than one father.

Seventhly. In young cats, which you are breeding to take prizes with, begin to look out for symptoms of the queen's getting gay, any time after six months, and on the first signs lock her up for a week, or until she becomes herself again. Do not think of breeding from a cat you mean for the show-bench until she is at least eighteen months old, else you will spoil her for size.

Some people fancy that to manage cats properly, and guide their breeding to the Tom you desire them to, is very difficult. I have not found it so. There is a little trouble, certainly, but you are amply rewarded, when you find on the birth of the kittens that you have been successful. The only thing you've got to do, is to watch the queen well, and lock her up for a night or two with her own lord in an outhouse.

Then afterwards keep her prisoner by herself for ten days. The danger is quite past then.

Eighthly. About a week before any important show, be more than usually careful with the grooming, etc, of your cats, and feed them up a bit; give them an extra allowance of milk and cream, and boiled rice and sugar, and occasionally mutton and mutton-broth, but take great care not to induce diarrhoea.

Ninthly. Send them to the show in a basket lined with flannel and a cushion, and pretty collar or ribbon to match the colour of the coat. Let the colour of the cushion be also effective, and in keeping with pussy's jacket.

As to cat-shows themselves, I have nothing but good to say. All prosperity to their promoters and patrons! They are in general, indeed almost invariably, well managed, and the cats are carefully caged, properly tended and fed, and no lady need apprehend the slightest danger to her feline favourite, in being sent to any of our great shows. It is seldom, if ever, that a cat is lost, the baskets containing the pussies never being opened, until inside the building, and then only with the greatest care. Indeed, one needs to be pretty cautious in handling a strange cat. Your well-bred beauties, in particular, make it a rule to stand no nonsense.

The cats are fed morning and night, and regularly supplied with the best and sweetest milk which the town can afford. Indeed, altogether, the poor things appear quite as happy as they are at their own firesides. If it is a four-day show, they soon come to know and welcome with gloved hand, the girl attendants every time they pass. There is no head-splitting noise and din as there is in a dog-show. Peace and quiet and serenity reign everywhere in a cat-show.

At nearly all the shows--at all events at all the *great* shows--Mr Sillet, the well-known naturalist of Southampton, has the arrangement of the pens or cages for the pussies. And very well he does his work too. Every cage is supplied with a box for sand at the back, and in the fore part with a beautiful soft cushion. The boxes are emptied daily, and disinfectants are also used, so that everything is sweet and clean. The entries at some of our national shows, such as the Crystal Palace and Birmingham, number between three and four hundred, and every year I trust the numbers will be increased.

You see then, reader, that no danger can accrue from sending your feline favourite to a show, and I may tell you also that if she is anything like good at all, she is almost sure of finding herself placed. Cat-shows are only in their infancy, and anyone who *chances* to have a good cat, may nowadays take prizes. In future years, there will be no chance work about the matter at all, and those only who study the breeding and rearing of cats in a scientific and sensible manner will be the winners.

When you send your entry form up to the secretary, be careful you have placed your pussy in the right class, not only as to breed but as to sex, whether male, female, or gelded. As to breed, you must attend to the colour and also to the length of the coat.

There are classes for all kinds of cats, and a class for anomalies besides.

I am often sorry, when judging at shows, to have to disqualify many a beautiful specimen of the feline race, because it has been carelessly entered in a wrong class. If people only will read with some degree of attention the description of each class, given in the schedules, they need never make this mistake.

To such clever and energetic managers of shows as Mr Wilson, of the Crystal Palace, who seems to have adopted the motto of the Cameron clan, "Whatever a man dares he can do," or sensible Mr Chaplin, of Birmingham, or Mr Brown, of Edinburgh, or Mr Martin, of Glasgow, I have positively nothing to suggest. Let anyone who wants to get up a cat-show take a lesson out of the books of either.

To amateur managers I may say this: Be very tender and gentle with the feline property entrusted to your care; remember not only that cats are extremely nervous and sensitive creatures, but also that numbers of them have a value in the eyes of their owners far above money and above price.

Feed with Spratt's Patent Cat Food. This ought to be used at all shows; it has the advantage of being cleanly, handy, and wholesome. A small allowance of boiled lights may be added.

Use chloride of lime, not too much of it, as a disinfectant.

Fill the utility boxes with plain garden mould or sand, but *never put charcoal in it*. That soils the fur, and doesn't give a white cat the chance of looking well.

Never put sawdust in a cat's cage. It gets into the milk and spoils it, and if they lick it it will make them ill.

Do not receive a cat that is suffering from illness of any sort.

If a cat should appear to be ill any time during the exhibition, have her carefully removed and sent home.

Finally, if possible, have beautifully ornamented prize cards, and send them home neat and clean to the successful exhibitors. These cards are greatly valued, and generally framed and hung in a conspicuous place.

No one, except the initiated, can have any idea what an important little creature a cat becomes that has once taken a prize. She is then more than ever the valued pet of her owners, and an object of interest even to the neighbours.

* * * * *

CHAPTER NINETEEN.

On Cruelty to Cats.

"He prayeth well, who loveth well,
Both man, and bird, and beast;
He prayeth best, who loveth best,
All things both great and small,
For the dear God who loveth us,
He made and loveth all."

Coleridge.

I am fond of cats, and am never happier than when I am writing about them; nevertheless, it is with feelings the very reverse of pleasant that I commence the present chapter. Were I to consult my own comfort, I should avoid the subject of cruelty to cats, and it is only with the hope, that I may be the means of doing some little good to poor harmless pussy, that I approach the matter at all.

I am not a sentimentalist by any means, yet I abominate wanton cruelty. I am fond of animals, yet not maudlinly so. I am not a vegetarian; and, although I neither believe that all animals were made for man's use, nor that man was made for theirs (as, you remember, was the opinion of the pampered goose), still I think we are right to kill and to use them as food. So I am fond of fishing, and fond too of shooting, and I can see nothing in the Bible against either practice. The very reverse, indeed, and everywhere in nature we observe that God permits one animal to prey upon another; and can the Lord Himself do wrong?

Yet, albeit I love sport and shooting, I do not think I am cruel. All my animals love me. My fishes know me, and come to be fed; my birds flutter their wings with affectionate excitement when I approach their cage; my white rats run to me when I call; my cat certainly never rushes up the chimney when I enter the room; and when I am dead I know my dogs will miss me.

Now, what I particularly object to is wanton and unnecessary cruelty. If we have to, and must, put the lower animals to death, in order that we--the higher animals--may live, we ought to do so as humanely as possible; and never, on any account, should we torture animals for mere sport. Hence I object to cock-fighting, pigeon or sparrow-shooting, and ratting--all mean and cowardly employments, and quite unfitted for men above the rank of the commonest navvy. I see no harm in deer-stalking in Scotland, where the deer are as wild as the hare or coney; but I do see very great cruelty in what is called stag-hunting in England. The stag in England is a domesticated animal, and I do not see that there is greater pluck or courage needed in hunting it, than there would be in chasing a decent old Alderney cow. I had travelled pretty nearly all over the world, and had shot in Africa, India, and Greenland, before I witnessed the first English stag-hunt. If my sympathies had not been all with the poor stag, I should have been highly amused indeed. The first stag wouldn't move at all; he looked upon the matter as too good a joke. "No, beggar me," he seemed to say, "if I'll budge an inch, to please anybody!" And he didn't. Yet this stag-hunting, they will tell you, seriously, keeps up the national courage. Believe me, reader, English courage requires no such keeping up, and it will be a poor day for this country when it does. Besides, it is only gentlemen (?) who hunt; and, well as our army is officered, it is, after all, the men who do the fighting; and it has always struck me that good beef and mutton, together with a determination to do their duty, are the mainstays on which our soldiers depend in the day of battle.

A great deal, I think, of the cruelty which is inflicted on the poor cat, is done through ignorance of pussy's nature and constitution; done unwittingly, and with no real intention of doing the animal an injury.

It is very cruel indeed to starve the creature, with the idea that you will induce her to catch more mice. When a cat is hungry the system is weak, the mind is dull, and the nerves so far from being well-strung that she will do anything sooner than hunt. A well-filled stomach gives pussy patience, and that is much wanted for mouse-killing; besides, you must not forget that cats kill mice as much for the sport as anything else.

Another very common form of cruelty is that of turning the cat out every night. Cats need their comforts, and enjoy them too, more than any other domestic animal we possess. Leaving her out at night not only exposes her to colds, inflammations, and various diseases, but it leads her to contract bad habits; and she eventually gets trapped or killed, and no wonder; is she not, through your carelessness, a nuisance to the whole neighbourhood?

It is cruel not to feed your cats with regularity. They expect it, and need it; and, if they do not get it, what else can you expect but that your cat will become a thief?

What is called "wandering" cats is extremely cruel. A man has no further use for his cat, so he "wanders" her. I assure you it would be far more humane to drown her at once. How would you, yourself, like to be wandered--to be taken abroad somewhere, and placed down in the centre of savages; hungry and cold, and longing and pining for the home you left behind you; and in danger every moment of being cruelly slain? Don't you think that speedy dissolution were more to be desired than such a life?

It is cruel, when your cat has kittens, to permit more to live than you can find decent homes for. It is a shame to a poor little kit, after it has opened its eyes to the wonders all around it, and begun to get happy and funny. Always keep one or two kittens for sake of the mother, and try, if possible, to find some one to take them. But the worst form of unintentional cruelty is that of leaving your poor favourite at home, when you go to the seaside, or to summer quarters. Often and often, on the return of the family, the unhappy cat is found lying in the empty hall, dead or dying, and wasted away to a mere handful of bones and skin--this in itself testifying to the sufferings she must have undergone for the want of food and water. Such gross *carelessness ought to be made penal*. I do not know whether the Society has ever yet prosecuted anyone for thus cruelly starving a cat, but I should think it would have little difficulty in obtaining a conviction.

I come now to mention some cases of intentional and specific cruelty, and shall be as brief as possible.

Some men, both young and old, think that a cat is a fit subject for torture and cruelty of all kinds; hence they never miss the chance of shying a stone after pussy's retreating figure. Cases, too, are continually cropping up in the police courts, of men having tortured cats to the death with dogs.

Cat skins are considered of some value by the furriers. At a sale not long since in London, there were some three thousand cat skins. Where think you, reader, do these come from? That is a question unfortunately only too easily answered. In almost all large cities there exists a gang of ruffians--you cannot call them by a milder name--who eke out a sort of livelihood by stealing cats by every available means and method. But worse than this remains to be told; it is darkly whispered, and I have some reason to believe it may be but too true, that many of those poor cats are *skinned alive*, in the belief that the living skin thus procured retains the gloss.

In Greenland I have seen young seals flayed alive by the score. That was a sickening sight enough, but skinning alive a poor harmless cat must be many times worse. I wish I could say that it was only the lowest class of ruffians that ill-treat poor cats to the death, but--and I know this for certain--there are men who pass as gentlemen, who night after night set traps for cats that stray into their gardens, and kill them in the cruellest manner; and some of these fellows, too, keep neither poultry, pigeons, nor rabbits, and haven't a flower in their gardens worthy of the name, only *they hate cats*. I know one gentleman (?) who thus traps and kills cats because he has a passion for fur rugs, which he thus indulges on the cheap.

Little boys, and those too, sometimes the sons of respectable parents who ought to have taught them better, are often dreadfully cruel to cats, stoning them wherever found, and setting dogs to worry them to death.

A lady, a friend of mine, once attracted by the heartrending cries of a cat, found two young fiends, with a pretty pussy tied in an apron, gouging its eyes out with a nail!

A common form of cruelty to cats, in some rural districts of England, is that of tying two of them together by the tails and hanging them over a rope or pole to fight to the death.

Such cases as that of cutting cats' tails off for wanton mischief, burning or boiling cats alive, though not unknown, I am happy to say are very rare.

Now, considering how very useful an animal a cat is, I think it is high time the law interfered to protect her from violence and ill-usage.

I should like to see a tax imposed upon all cats, and a home for lost cats, precisely on the same principles as the home for lost and starving dogs, only with this difference, that there should be no reward offered for bringing a cat to the home. Remember this, that a stranger or starving cat will come to anyone who says a kind word to it, so policemen would have no difficulty in catching them.

The revenue from the imposition of even a small tax would be very large, and it would not only help to clear the country of a whole army corps of thieving, prowling, homeless cats, but give to the cats of respectable people a greater value in the eyes of the law, and a greater chance of taking their walks abroad without being molested.

We have a law to protect even our wild birds, why not one for the protection of my friend the harmless, useful cat?

In conclusion, let me assure lovers and owners of cats, that, as the law stands at present, the only way to keep their favourites alive, and free from danger, is to be kind to them, feed them well and teach them, as far as possible, to keep to the house at night.

We think that men who kill, and trap, and injure our cats are exceedingly cruel. And so they are, and I hope they will in time learn to be a shade more merciful. At the same time, don't forget that the temptation to take revenge upon a cat for vines destroyed, beautiful flowerbeds torn up, favourite rabbits murdered in their hutches, and valuable pigeons torn and eaten in their dovecots, is a very great temptation indeed. You see, reader, there are two sides to every question.

Pray think of the matter.

* * * * *

CHAPTER TWENTY.

Pussy's Tricks and Manners.

When I was a boy, it used to be a positive pain to me to have to enter a large library and choose a book. I used to wander round and round the well-filled shelves like a butterfly floating over a clover-field. I didn't know where to alight. I would fain have begun at the beginning, and read the lot--but that was impracticable. Hence my difficulty. I am in a somewhat similar fix now. I have so many original anecdotes of cat life and customs that I don't know which to tell.

If I had space at command you should have the whole lot, and I would arrange them into classes according to their character; as it is, I must be content to present the reader with some account of a few of pussy's tricks and manners, deduced from these and from my own rather large experience of cat life.

Every child knows how fond cats are of hunting and catching mice, but no cat of any respectability would think of confining her attentions to mice alone. The very presence of a cat about a house will usually suffice to keep these destructive pests at bay; and if one should pop out of its hole, it knows, or ought to know, what to expect. But seldom will a high-bred cat condescend to eat a mouse. She will play with it as long as hope keeps up its little heart; when that fails it, pussy turns it over once or twice to see whether it is really dead or only shamming, and then walks disdainfully away. The next higher game is rats, but these she seldom cares to eat, only she kills them on the spot. She knows that rats have teeth and can use them, so she doesn't romp with them. I have known rats inflict such severe wounds upon a cat that they ultimately proved fatal.

Cats delight to spend a day in the woods, bird-catching. They rob the nests, too, when they find any, and cases have occurred of a cat paying visits to nests day after day until the young were hatched, then eating them. (I once had a blackbird's nest in the side of a bank at the roadside--a strange place for a blackbird to build. I often used to see a polecat close to, and I am convinced it knew of the nest, but it never robbed it until the young were hatched.)

Nearly all cats who live in the country hunt over the hills and the woods, and a great plague, too, gamekeepers find them. There is no animal which a cat may meet in the covers that she is not a match for.

Polecats and weasels have to own her sway, while rabbits and leverets fall an easy prey to her prowess.

Most cats, who are well treated by their owners, have a habit of bringing everything home which they catch. I have often seen a cat come trotting homewards, carrying in its mouth a rabbit well-nigh as big as herself.

Cats may therefore be called poachers; and it is curious, but true, that when a poor man owns a cat who poaches, and brings home the quarry, he usually winks at it.

I have dozens of well-authenticated anecdotes of cats who are very expert at fishing. I have, myself, watched a cat by the banks of a stream, until I have seen him dive into the water, and emerge almost immediately with a large trout in his mouth. Cats who fish, generally belong to millers, or are bred and reared somewhere near a river. They not only catch fish of all sorts, but even water-rats; often springing many feet off the bank after prey of this kind, and even diving under to secure it. In Scotland cats often attack and destroy large quantities of salmon in small streams, in the spawning season.

Cats are supposed to have an antipathy to water, and, as a rule, this is so. They are very cleanly animals, and it has often amused me to watch a pussy crossing a muddy street. How eagerly she looks out for the dry spots, how gingerly she picks her steps, and, when she does tread in a pool, with what an air of supreme disgust she stops and shakes the offending foot!

Cats swim well, nevertheless. I have seen a cat take the water as coolly as an Irish spaniel, swim the river, hunt in the woods for some time, and then swim back again with a bird in her mouth. And, to save their kittens from drowning, almost any cat will swim a long distance.

I have known a cat whose favourite fish was the eel, and he always managed to catch one somehow.

Cats are very fanciful at times, and very self-opinionated. If a cat takes a fancy to a particular house, or part of the house, it is difficult to dislodge her.

"In the year 1852," a lady writes me, "my mother was living with a family in the Albany Road, Camberwell, who had a large tabby Tom cat. This cat had formed a strong attachment to a kitten who belonged to the lady next door. In 1853, the family removed to the Ashby Road, Lower Road, Islington, and the cat was *packed in a hamper*, and sent with the furniture.

"It was kept in confinement the first day and night, and let out the next morning. Tabby had his feet buttered, to keep him employed, as they said it was a good thing to keep him busy. The next day he had disappeared, no one knew whither, though search was made for him everywhere.

"A few days after, the lady from Camberwell wrote to say that Tabby had put in an appearance there, and resumed the charge of his kitten. He was sent back by the carrier to his proper owner, and every means was tried to induce him to stop; but he returned the second time to the kitten, and so they let him remain, because they knew he would be well taken care of. The wonderment of this was: *which bridge did he go over in passing through busy London*?"

It is really wonderful how a cat can often find its way, long distances across a country which he never before may have traversed.

"A few days ago," says another correspondent, "a lady who lives in Newport told me that, at one time, her house was quite overrun with mice; and, having procured the loan of a cat which was considered a good mouser, she tied it into a basket, and then placed it in a concealed part of the pony carriage. On her arrival at the 'Cliff' the prisoner was released; but even the prospect of a delicious feast of mice could not obliterate its thoughts of 'home, sweet home;' and, after about an hour's stay, it set off, and, ere long, arrived at its former abode--distant three miles!"

Some months ago, a half-bred Persian tabby, came to my place, and has since then stuck to it with all the persistency of Edgar Allan Poe's raven. He is a cat that seems to have nothing to recommend him; if he would come into the house, and behave like a civilised being, I would never grudge him his daily dole. But he prefers to live a half-pagan existence, out among the bushes, and take his nap of a night in the coal-house; and Bridget says he is an awful thief, and that she can't leave the kitchen-door open one moment for fear of him. I've often asked that cat to take his departure, but, as plain as cat can speak, that cat says "never more."

By way of experiment I have caught him several times--no easy task, I assure you--and *sent him*, securely packed in a hamper, distances of three, four, and five miles to friends who have set him free. And he always came back. His last journey was at Christmas-time--may Heaven forgive me this sin!--to the house of a parson *whom I did not know*, and I stuck some pheasants' feathers too just under the lid. I don't know what the parson thought, but Tom came back next day, not looking a single bit put out, and--I am willing to sell him to anyone who may have need of his services.

I know a cat who caught two sparrows at once, and when retreating, a third sparrow pursued and attacked him. This one pussy also killed, with his paw. That was funny!

Cats know certain days of the week, such as Sunday for instance, and they also know certain hours of each day. I don't mean to say they look at the clock, but, if a favourite master or mistress is in the habit of coming home every day, say at 4 p.m., there you will often find that every day at 4 p.m. pussy will trot down the road to meet her and wait till she comes.

Cats make good husbands, gentle fathers, and the most tender and loving of mothers. A cat will fight for her kittens, starve or *steal* for them. Oh! I daresay you imagine that stealing wouldn't be likely to lie very heavily on a cat's conscience. Now listen to this--which the printer will kindly put in italics--*all experience goes to prove that well-fed, properly cared-for cats, are not thieves, but the reverse.*

Cats have their kittens in queer places, at times. A lady's best Sunday bonnet, or master's wig, or a set of ermine furs, just suits pussy to a nicety. My cat once kittened in my cocked hat. It is a positive fact, madam, and so far from thinking she had done anything to offend me, she held up one of her brats for me to admire. But the queerest place for a cat to kitten in, that ever I knew, was a tree. The cat scrambled up the tree and brought forth her young in the nest of a wood-pigeon! I didn't hear how the kittens got down again though, but I have every reason to believe the story. Probably, when the kittens opened their eyes they commenced playing with their mother's tail, and went topsy-turvy to the ground. Well, *facilis descensus Averni,* and you know cats always fall on their feet. I knew a man who kicked his own cat out of his pigeon loft, three storeys high. He told me it didn't seem to hurt her a bit, but rather increased her appetite.

Whether cats have nine lives or not, they take a great deal of killing.

I knew a cat that was drowned four times, and came home again as unconcernedly as if nothing very unusual had happened. However, drowning in the end seemed to get rather irksome to this pussy, and after the fourth immersion, he ran away to the woods, and didn't come back to be drowned any more.

Many cases I know of parties having started off with puss in a bag to drown her, and having stopped to talk to a friend on the way back found, on their return, the cat sitting by the fire drying herself! I have many instances of cats having been thrown from bridges and other high places, with the intention of killing them, but without fatal effect.

Cats have been buried alive for days and recovered after being dug up. A cat of my acquaintance was sent to live at a mill. This seemed to please pussy very much. You see there were plenty of mice in the mill, and plenty of rats and fish in the mill-lead, so the cat made herself at home. But in course of time pussy became the mother of two kittens, and then the longing for her old home came back with a force too powerful to be resisted. She determined, therefore, to return to her former residence, and she did so, carrying her kittens one by one. The distance she had to travel was two miles, and the night she chose was a dark and stormy one.

There were two cats who dwelt at the self-same house and had kittens at the self-same time. All the kittens were drowned with the exception of two, one being left with each mother. And now comes the curious part of the business. These two mother-cats came to an amicable understanding, that whenever the one was abroad the other should suckle and attend to both babies, and this treaty was carried out to the letter.

Cats are not only fond of human beings, but often get greatly attached to other domestic animals, especially to the family dog. I know at this moment a cat whose constant companion is a Dandy Dinmont; and a rough one he is too, for, although he sleeps in pussy's arms every night, he thinks nothing of pulling her all round the lawn by the tail at any time, the cat herself seeming to enjoy the fun!

Rabbits and cats often associate together on the most friendly terms, even accompanying each other in long excursions, the cat on these occasions electing herself protector of her feebler friend against predatory dogs and other cats.

A cat belonging to a friend of mine used to be constantly at war with the dog, until one day, with a blow of her ungloved paw, she blinded the poor animal in one eye. No mother could have been kinder to her child than pussy was to this dog, after she saw what she had done. That she bitterly repented the rash act is evident, for she watched beside him night and day, until he grew well again; and now, they are the fastest friends in the world, and the cat is the first to welcome the dog home when he returns from a walk.

As a proof of how cruel it is to take *all* a cat's kittens away from her, I may state that, thus bereaved, a cat will take to nursing even chickens, or she will suckle puppies, hedgehogs, or rats.

It is a funny thing that many cats can't bear music. Some will run out of the room if they hear a fiddle played, and others will growl and attack the musician.

Cats can be easily taught to follow one in a country walk just like a dog, and on these occasions they come much better to the sound of whistling than to any other call.

A well-bred cat will always teach its kittens habits of cleanliness, how to watch for and catch mice, and also how to catch minnows in a shallow stream.

I have already said that cats, as a rule, when well treated, are not thieves, but the very reverse. But when a cat does take to thieving for a livelihood, she becomes quite a swell at it--shows how clever she is.

Cats are considered in some parts of England to be of some value as an article of diet. I have never to my knowledge eaten cat, so I cannot give the reader any idea what they taste like.

It is ridiculous to suppose, as some do, that a cat's breath has any effect upon a baby either for good or for evil. Neither will a cat bring blood from a child's temple by licking it with its rough tongue.

An ugly old woman isn't necessarily a witch because she keeps a black cat. Neither is a black cat a devil.

They say that witches sail over the sea in riddles accompanied by their black cats, and that they have rather a jolly time of it upon the whole, having plenty to eat, and plenty to drink--flagons of wine, in fact. Don't you believe it, reader.

Cats are not afraid of snakes; but snakes, even the dreaded cobra, will invariably give pussy a wide berth.

Cats are fond of fish, absurdly so, and if you offer them even the gold-fish, they won't feel offended. It is only out of respect for the owner thereof that they don't devour the canary. They prefer canary living, with the feathers on. It tickles their palates and makes them laugh.

Chickens are dainties in a cat's *cuisine*; they also rather like a nice plump partridge, and won't refuse to suck an egg when occasion offers.

Cats are, as a rule, Good Templars; the proof of which rule is this: I had a Red Tabby Tom who would eat oatmeal and whisky until he couldn't stand. The servants knew this failing, and encouraged him in his evil ways; so that half his time, instead of being as sober as a judge--as every decent, respectable cat ought--Tom was as drunk as a piper.

It is funny to listen to a cat's concert about two o'clock in the morning. Of course, if you are rather nervous, and want to go to sleep, it isn't so funny. (N.B.--If cats were better treated, they would hold their concerts in daylight in the garden, instead of at midnight on the tiles. Mind you, there is something in that.)

Altogether, cats are funny things, and the more you study them the funnier you find them. That's so!

* * * * *

CHAPTER TWENTY ONE.

The Fireside Favourite.

The lines of some cats fall in pleasant places. Mine have. I'm the fireside favourite, I'm the parlour pet. I'm the *beau ideal,* so my mistress says, of what every decent, respectable, well-trained cat ought to be--and I looked in the glass and found it so. But pray don't think that I am vain because I happen to know the usages of polite society, and the uses and abuses of the looking-glass. No cat, in my opinion, with any claim to the dignity of lady-puss, would think of washing her face unless in front of a plate-glass mirror. But I will not soon forget the day I first knew what a looking-glass meant. I was then only a cheeky little mite of a kitten, of a highly inquiring turn of mind. Well, one evening my young mistress was going to a ball, and before she went she spent about three hours in her dressing-room, doing something, and then she came down to the parlour, looking more like an angel than ever I had seen her. Oh, how she was dressed, to be sure! And she had little bunches of flowers stuck on all over her dress, and I wanted to play at "mousies" with them; but she wouldn't wait, she just kissed me and bade me be a good kitten and not run up the curtains, and then off she went. Yes; I meant to be an awfully good little kitten--but first and foremost I meant to see the interior of that mysterious room. By good luck the door was ajar, so in I popped at once, and made direct for the table. Such a display of beautiful things I had never seen before. I didn't know what they all meant then, but I do now, for, mind you, I will soon be twenty years of age. But I got great fun on that table. I tried the gold rings on my nose, and the earrings on my toes, and I knocked off the lid of a powder-box, and scattered the crimson contents all abroad. Then I had a fearful battle with a puff which I unearthed from another box. During the fight a bottle of ylang-ylang went down. I didn't care a dump. Crash went a bottle of fragrant floriline next. I regarded it not. I fought the puff till it took refuge on the floor. Then I paused, wondering what I should do next, when behold! right in front of me and looking through a square of glass, and apparently wondering what *it* should do next, was the ugliest little wretch of a kitten ever you saw in your life--a long-nosed, blear-eyed, pingey-wingey thing. I marched up to it as brave as a button, and it had the audacity to come and meet me.

"You ugly, deformed little beast," I cried, "what do you want in my lady's room?"

"The same to you," it seemed to say, "and many of them."

"For two pins," I continued, "I would scratch your nasty little eyes out--yah--fuss-s!"

"Yah--fuss-s!" replied the foe, lifting its left paw as I lifted my right.

This was too much. I crept round the corner to give her a cuff. She wasn't there! I came back, and there she was as brazen as ever. I tried this game on several times, but couldn't catch her. "Then," says I, "you'll have it where you stand, and hang the pane of glass!"

I struck straight from the shoulder, and with a will too. Down went the glass, and I found I had been fighting all the time with my own shadow. Funny, wasn't it?

When mistress came home there was such a row. But she was sensible, and didn't beat me. She took me upstairs, and showed me what I had done, and looked so vexed that I was sorry too. "It is my own fault, though," she said; "I ought to have shut the door."

She presented me with a looking-glass soon after this, and it is quite surprising how my opinion of that strange kitten in the mirror altered after that. I thought now I had never seen such a lovely thing, and I was never tired looking at it. No more I had. But first impressions *are* so erroneous, you know.

My dear mother is dead and gone years ago--of course, considering my age, you won't marvel at that; and my young mistress is married long, long ago, and has a grown family, who are all as kind as kind can be to old Tom, as they facetiously call me. And so they were to my mother, who, I may tell you, was only three days in her last illness, and gave up the ghost on a file of old newspapers (than which nothing makes a better bed) and is buried under the old pear-tree.

Dear me, how often I have wondered how other poor cats who have neither kind master nor mistress manage to live. But, the poor creatures, they are *so* ignorant--badly-bred, you know. Why, only the other day the young master brought home a poor little cat, he had found starving in the street. Well, I never in all my life saw such an ill-mannered, rude little wretch, for no sooner had it got itself stuffed with the best fare in the house, than it made a deliberate attempt to steal the canary. There was gratitude for you! Now, mind, I don't say that *I* shouldn't like to eat the canary, but I never have taken our own birds--no--always the neighbours'. I did, just once, fly at our own canary's cage when I was quite a wee cat, and didn't know any better. And what do you think my mistress did? Why, she took the bird out of the cage and popped me in; and there I was, all day long, a prisoner, with nothing for dinner but seeds and water, and the canary flying about the room and doing what it liked, even helping itself to my milk. I never forgot that.

Some cats, you know, are arrant thieves, and I don't wonder at it, the way they are kicked and cuffed about, put out all night, and never offered food or water. I would steal myself if I were used like that, wouldn't you, madam? But I have my two meals a day, regularly; and I have a nice double saucer, which stands beside my mirror, and one end contains nice milk and the other clean water, and I don't know which I like the best. When I am downright thirsty, the water is so nice; but at times I am hungry and thirsty both, if you can understand me--then I drink the milk. At times I am allowed to sit on the table when my mistress is at breakfast, and I often put out my paw, ever so gently, and help myself to a morsel from her plate; but I wouldn't do it when she isn't looking. The other day I took a fancy to a nice smelt, and I just went and told my mistress and led her to the kitchen, and I got what I wanted at once.

I am never put out at night. I have always the softest and warmest of beds, and in winter, towards morning, when the fire goes out, I go upstairs and creep (singing loudly to let her know it is I) into my mistress's arms.

If I want to go on the tiles any night, I have only to ask. A fellow does want to go on the tiles now and then, doesn't he? Oh, it is a jolly thing, is a night on the tiles! One of these days I may give you my experience of life on the tiles, and then you'll know all about it--in the meantime, madam, you may try it yourself. Let it be moonlight, and be cautious, you know, for, as you have only two feet, you will feel rather awkward at first.

Did I ever know what it was to be hungry? Yes, indeed, once I did; and I'm now going to tell you of the saddest experience in all my long life. You see it happened like this. It was autumn; I was then about five years of age, and a finer-looking Tom, I could see by my mirror, never trod on four legs. For some days I had observed an unusual bustle both upstairs and downstairs. The servants, especially, seemed all off their heads, and did nothing but open doors and shut them, and nail up things in large boxes, and drink beer and eat cold meat whenever they stood on end. What was up, I wondered? Went and asked my mistress. "Off to the seaside, pussy Tom," said she; "and you're going too, if you're good." I determined to be good, and not make faces at the canary. But one night I had been out rather late at a cat-concert, and, as usual, came home with the milk in the morning. In order to make sure of a good sleep I went upstairs to an unused attic, as was my wont, and fell asleep on an old pillow. How long I slept I shall never know, but it must have been far on in the day when I awoke, feeling hungry enough to eat a hunter. As I trotted downstairs the first thing that alarmed me was the unusual stillness. I mewed, and a thousand echoes seemed to mock me. The ticking of the old clock on the stairs had never sounded to me so loud and clear before. I went, one by one, into every room. Nothing in any of them but the stillness, apparently, of death and desolation. The blinds were all down, and I could even hear the mice nibbling behind the wainscot.

My heart felt like a great cold lump of lead, as the sad truth flashed upon my mind--my kind mistress had gone, with all the family, and I was left, forgotten, deserted! My first endeavour was to find my way out. Had I succeeded, even then I would have found my mistress, for cats have an instinct you little wot of. But every door and window was fastened, and there wasn't a hole left which a rat could have crept through.

What nights and days of misery followed!--it makes me shudder to think of them even now.

For the first few days I did not suffer much from hunger. There were crumbs left by the servants, and occasionally a mouse crept out from the kitchen fender, and I had that. But by the fifth day the crumbs had all gone, and with them the mice, too, had disappeared. They nibbled no more in the cupboard nor behind the wainscot; and as the clock had run down there wasn't a sound in the old house by night or by day. I now began to suffer both from hunger and thirst. I spent my time either mewing piteously at the hall-door, or roaming purposelessly through the empty house, or watching, watching, faint and wearily, for the mice that never came. Perhaps the most bitter part of my sufferings just then was the thought that would keep obtruding itself on my mind, that for all the love with which I had loved my mistress, and the faithfulness with which I had served her, she had gone away, and left me to die all alone in the deserted house. Me, too, who would have laid down my life to please her had she only stayed near me.

How slowly the time dragged on--how long and dreary the days, how terrible the nights! Perhaps it was when I was at my very worst, that I happened to be standing close by my empty saucer, and in front of my mirror. At that time I was almost too weak to walk, I tottered on my feet, and my head swam and moved from side to side when I tried to look at anything. Suddenly I started. Could that wild, attenuated image in the mirror be my reflection? How it glared upon me from its glassy eyes! And now I knew it could not be mine, but some dreadful thing sent to torture me. For as I gazed it uttered a yell--mournful, prolonged, unearthly--and dashed at me through and out from the mirror. For some time we seemed to writhe together in agony on the carpet. Then up again we started, the mirror-fiend and I. "Follow me fast!" it seemed to cry, and I was impelled to follow. Wherever it was, there was I. How it tore up and down the house, yelling as it went and tearing everything in its way! How it rushed half up the chimney, and was dashed back again by invisible hands! How it flung itself, half-blind and bleeding, at the Venetian blinds, and how madly it tried again to escape into the mirror and shivered the glass! Then mills began in my head--mills and machinery--and the roar of running waters. Then I found myself walking all alone in a green and beautiful meadow, with a blue sky overhead and birds and butterflies all about, a cool breeze fanning my brow, and, better than all, *water*, pure, and clear, and cool, meandering over brown smooth pebbles, beside which the minnows chased the sunbeams. And I drank--and slept.

When I awoke, I found myself lying on the mat in the hall, and the sunlight shimmering in through the stained glass, and falling in patches of green and crimson on the floor. Very cold now, but quiet and sensible. There was a large hole in my side, and blood was all about, so I must have, in my delirium, *torn the flesh, from my own ribs and devoured it*. (Not overdrawn. A case of the kind actually occurred some years ago in the new town of Edinburgh.--The Author.)

I knew now that death was come, and would set me free at last.

Then the noise of wheels in my ears, and the sound of human voices; then a blank; and then someone pouring something down my throat; and I opened my eyes and beheld my dear young mistress. How she was weeping! The sight of her sorrow would have melted your heart. "Oh, pussy, pussy, do not die!" she was crying.

Pussy didn't die; but till this day I believe it was only to please my dear mistress I crept back again to life and love.

I'm very old now, and my thoughts dwell mostly in the past, and I like a cheery fire and a drop of warm milk better than ever. But I have all my faculties and all my comforts. We have other cats in the house, but I never feel jealous, for my mistress, look you, loves me better than all the cats in the kingdom--fact--she told me so.

CHAPTER TWENTY TWO.

The Dunghill Cat.

I'm the dunghill cat--that is what I am. Nobody owns me, and I owe allegiance to nobody. Nobody feeds me; nobody puts a saucer on the ground and says, "Here, pussy, there's a drop of milk for you, my pet." Nobody ever gave me a bit of fish in my life, and nobody, so far as I can remember, ever called me pet names or spoke kindly to me. Not that I care, you know, but I merely mention it, that's all. But don't you despise me because I am only a poor dunghill cat. It isn't my fault but my misfortune, as you shall presently hear. Circumstances over which I had no control have rendered me what I am; but I am come of respectable parents for all that. To be sure I could not swear to my father, not knowing exactly who he was, and the mum herself being at times a little hazy on the point. But my mother, madam, came from Egypt, and was descended from a long line of noble ancestors in that beautiful land, where, they tell me, there is bread enough for all, and where a poor cat is honoured and respected, as she always ought to be. And the mum told me that her original ancestors came over with the Conqueror--Cambyses, you know--so that is good enough, surely. Yes, madam, without meaning the slightest offence, I may just remind you that when your forebears were dressed in pig-skins, and not much of that; when they wore flint-headed spears, and stalked about the hills with painted faces, doing attitudes and saying "Ugh!" when astonished, my progenitors dwelt in palaces, loved and respected by all, and were considered the equals of prince, or priest, or peer--what do you think of that? But I'm not proud; I'm only the poor dunghill cat, that all the dogs chase, that all the little boys stone, and Bridget shakes the broom at. Bridget never can catch me, though--ha, ha! Won't I eat her canary, first chance--you see if I don't.

My earliest recollection is of being carried by the back of my neck, by something or somebody that I afterwards discovered was my mother. I was taken into a beautiful house, and deposited carefully on a rug in the corner of a cupboard. Then my mother began licking me all over with her tongue, when suddenly said a voice close alongside of me, "I declare that pussy has been and gone and got another kitten--as if one cat of the kind wasn't enough about the house. Sarah, go and put it where you put all the others."

I don't know who the others were, or where they were put; but I know what Sarah did with me. She took me up with the hot tongs, mother screamed and so did I, till I couldn't scream any more because the black water was all around me. Then followed a period of agony, and then a blank, and the next thing I recollect is finding myself lying, wet and cold, in my mother's arms, and she all wet and cold as well as me.

"My dear chee-ild," said my mamma, "this has been a sad morning; but you're safe ne-ow, although the building is humble and your pallet is straw. Shade of Cambyses!" continued the old lady, rubbing a paw over her right ear, "why ever did I leave the land of Egypt?"

When I got a little older I began to look around me. I thought our new home was one of the jolliest places that could be, despite all the flowery accounts my mother used to give me of the land of her birth, with its marble halls and gorgeous tesselated pavements. It was a large, roomy loft in an old, old mill, and I used to run about the floor and chase the great spiders before I was big and brave enough to attack a wild mouse, or the great, untamable rats that used to frighten me so when mother was out, by standing on their hind legs and making dreadful faces at me. But didn't they scamper off when mother came back!

One day mother brought me a live mouse. How brave I suddenly felt. You should have seen how I sprung on it, and heard how I growled. Had anyone, even the immortal Cambyses himself, attempted to rescue that wild mouse from my clutches, he should have died on the spot. How pleased my mother looked! I think I see her yet, with her old-fashioned face and her odd, old-world ways. Very much respected my mother was, I assure you. I've seen no less than seven well-dressed feline swells talking and singing to her all at once, and she didn't know which of them to speak to first. Met a violent end, did my mother. Verdict--"Killed by the carrier's collie."

After I had slain and eaten one mouse, I felt every inch a Tom. I declined to lie any more in my mother's arms. No more milk for me; blood, and only blood, was my motto, and I meant it, too. When I was a well-grown cat of nine months old my mother introduced me to her mistress's house, and I became, for a time, a house-cat. I cannot say, however, that I liked the change. The lady of the dwelling was, they told me, exceedingly good and pious, went twice to church on Sunday, and read prayers morning and evening; but, sad to say, she never had studied feline economy. "If cats can't find mice to eat," she used to say, "they ought to starve."

My mother told me that this was something like asking a person to make bricks without straw. My mother was very learned.

Well, one evening--and I had been starving all day, and was dreadfully hungry and too faint to watch for mice--I happened to stroll into the pantry, and there I found such a nice, nice dish of cream. Luscious! But what a thrashing I got five minutes afterwards--I wasn't hungry for a week. Then the hunger came on again worse than ever, and I stole again. I couldn't help it, really. Then I was called a nasty, thieving brute, and got blamed many times when quite innocent. There is Briddy with the broom again. She hasn't forgiven me for that herring yet, and I can swear it wouldn't have kept for another day. Besides, what do I care if it was for Master Fred's breakfast? Briddy had no business to be upstairs trying on missus's Sunday bonnet, and the kitchen-door wide open. She thinks I don't see all her capers, and her opening drawers, and keeking into cupboards, and examining this, that, and t'other, when her missus is out. But lying on the top of that wall I can see a great deal more than I trouble to tell of. But Briddy blamed me for eating those two new-laid eggs that the baker brought. She "just laid them down outside in the strawberry-basket, m'm, for one minute; and when she turned again, la, m'm, they was broke and eaten, they was!" She forgot to mention how the baker crumpled her cap, though; and she didn't tell how she was all over flour, and had to brush herself from top to toe when the bell rang. But, mind you, it wasn't *me* that stole the eggs. I would confess at once if it was; for what could a couple of paltry new-laid eggs add to the weight of crime I have been guilty of in my day? Why, nothing. But Dr Ricket's jackdaw took the eggs, for I saw him hop on to the wall, and he gave a look down, first, with one side of his head, at Briddy and the baker, then, with the other side of his head, to the eggs; then down he went, and it was all over in a moment--I mean the eggs were. Just like Briddy, blaming me for that piece of cold pork. Mind you, I don't say I wouldn't have taken it had I got the chance, but I didn't. "That beautiful piece of pork gone next, m'm; and I never can keep that cat out. And whatever shall I do, m'm?"

But I wonder why Briddy didn't say a word about that visit she had from the policeman. Much of a lover he is, anyhow. I could see him through the window, and he never opened his mouth but to put something into it. His courtship was *so* un-Byronic, for he sat and he sat, and he chewed and chewed, and glowered and glowered at Briddy, till I wondered she didn't spit in his face and turn him out. Ah, Briddy, you needn't shake the broom, what would you do without me?

But to resume my story. One night I was shut up in a room by accident, and no one heard me call, for I did call, and, in the morning, the room wasn't just as it ought to have been, and for this new offence I was condemned to die--taken away in a sack, and drowned.

Not dead? Bless you, no; it wasn't likely I was going to remain at the bottom of a mill-dam, in an old guano-bag. I was up again before you could say mouse, and had swam on shore as cool as you like. It was a beautiful day in early autumn, the fields were all ablaze with golden grain, and the berries beginning to turn red and black in the hedgerows. I sat down on a sheaf of wheat and basked till dry in the warm sunshine. Then a young pheasant ran round the corner and cried, "Peet, peet, have you seen my mother anywhere?" I thought I never had tasted anything half so sweet in all my life. Then I felt a new Tom from top to toe. Go back and be a house-cat? No, perish the thought. And I never did.

I am now fifteen years of age, and as I look back to the days that are gone I cannot help exclaiming, "What a jolly life I've led." I've been a Bohemian, a robber, a brigand, and a thief. "It is a sin, pussy," you say; "why don't you reform?" "'Cause I won't," I answer. Had I been differently brought up, better treated, better fed, and better understood, I mightn't be what I am. I would then have been as honest and virtuous as one of good Mrs Peek's cats. She knows how to treat a cat, and it is only a pity she isn't an Egyptian, she might have married Cambyses.

Well, well, as I said before, I'm now fifteen years of age; I've seen many ups and downs in the world, but I suppose my day is wearing through, and I must soon be preparing for the happy hunting-fields on the other side of Jordan.

Now, madam, you know I'm only a cat, a common dunghill cat, and have only common dunghill notions, but here are my sentiments. Religion is a beautiful thing when brought to bear on everyday life, and not put off and on with your moire antique. But never you go away to church and forget to give pussy her breakfast.

And have your prayer-book in one hand if you like of a morning, but have a nice bit of fish or a saucer of milk for pussy in the other, and the beauty of the one hand will be reflected from the other, as the stars are mirrored in the ocean's wave.

THE END.

Made in the USA
Monee, IL
08 November 2026

68030006R00057